Arthur Shadwell

The Tallerman treatment by superheated dry air in rheumatism,

gout, rheumatic arthritis, stiff and painful joints, sprains, sciatica, and other affections

Arthur Shadwell

The Tallerman treatment by superheated dry air in rheumatism,
gout, rheumatic arthritis, stiff and painful joints, sprains, sciatica, and other affections

ISBN/EAN: 9783337414382

Printed in Europe, USA, Canada, Australia, Japan

Cover: Foto ©berggeist007 / pixelio.de

More available books at **www.hansebooks.com**

THE TALLERMAN TREATMENT

BY

SUPERHEATED DRY AIR

IN

Rheumatism, Gout, Rheumatic Arthritis, Stiff and
Painful Joints, Sprains, Sciatica, and
other Affections.

*CASE NOTES AND MEDICAL REPORTS, WITH
NUMEROUS ILLUSTRATIONS.*

EDITED BY

ARTHUR SHADWELL, M.A., M.B. Oxon., M.R.C.P.

LONDON:
BAILLIÈRE, TINDALL AND COX,
KING WILLIAM STREET, STRAND.
1898.
[All rights reserved.]

PREFACE.

The object of this book is to present in a convenient form full and authoritative information respecting the Tallerman treatment, which has been in use now for some four or five years in various countries, but is still very imperfectly known. When requested to supervise the preparation of the volume, I readily consented, for three reasons. In the first place, experience has convinced me of the value of the treatment; in the second, I think it ought to be very much better known than it is; and in the third, I have no personal interest in it whatever. I originally approached Mr. Tallerman's invention with the scepticism which becomes second nature to medical men, but having tested it on my own *corpus vile*, I found that it did what it pretended to do; and then I saw a boy with a knee-joint full of fluid and wincing at every movement gradually charmed off within half an hour into a smiling and painless indifference, which permitted the free handling and flexion of the limb without a murmur. Since then I have repeatedly seen results produced in old and hopeless cases of rheumatic arthritis which I could not have believed on any lesser evidence than my own eyesight. The facts related in this volume amply corroborate my experience, and make it unnecessary for me to say anything more on that head. Attested as they are by many independent observers of high standing in the profession, they form a body of evidence which no one can affect to ignore or despise. They do not come from one or from a few *cliniques*, but from a large number of first-rate hospitals, not only in this country, but in Paris, the United States, and in Canada. It is impossible to deny the weight of so large a mass of concurrent testimony.

In particular I may refer to the conclusions recently formulated by Professor James Stewart, of Montreal, Dr. Dejérine, of the Salpêtrière, and Professor Landouzy, of the Laennec Hospital, on the marked superiority of this treatment over any other hitherto used in various intractable forms of rheumatism.

The value of the invention, however, appears to be still very inadequately recognised. In some quarters it has met with liberal appreciation from the profession, as the medical reports here collected abundantly prove, but in others it has encountered indifference and opposition. Indifference is due to a general distrust of specious novelties, which is healthy enough when not carried too far. But opposition implies prejudice, and that is another matter. It seems to be due to the fact that the appliance is patented by a layman, and is therefore a commercial article. I fail to see what this has to do with the matter. Every appliance and every drug is a commercial article, and at least half of those in daily use are patented by laymen. What of antiseptics and thermometers and stethoscopes? The sole question is whether a thing is useful, not whether somebody or other has a pecuniary interest in it. Somebody has a pecuniary interest in everything used, if it is only a bandage or a bottle. And in justice to the inventor in this case, it ought to be known that he has brought forward his apparatus in the most legitimate manner, and at very great cost to himself. He has placed it in the hands of the medical profession, and worked entirely with them from first to last, refraining from all popular advertisement. With the most scrupulous good faith, he has repeatedly refused lucrative and influential clients, who wished to be treated, but declined to see a doctor. Anything further removed from quackery could not be imagined. He has not always received equally honourable treatment in return.

It is hoped that this explanation, together with the proofs of the real value of Mr. Tallerman's invention, will lead to its extended use. Putting aside other instances of its successful application—of which full particulars will be found in this book—it has been proved beyond all question capable of affording marked relief in those saddest of all cases in which every other known remedy has failed. To permit these unfortunates—on account of a baseless prejudice—to go their painful way to a hopeless end, when a fair chance of materially ameliorating their condition still remains, would be so opposed to all the best traditions of the medical profession, that I cannot conceive its happening when once the facts are known.

I should add that I am merely responsible for the arrangement of the medical sections of the book, a great part of which has already appeared in various professional journals.

A. SHADWELL.

CONTENTS.

	PAGE
PREFACE	v-vii

CHAPTER I.
INTRODUCTORY.

The local application of heat—Failure of previous attempts to obtain high temperatures—Description of the Tallerman apparatus—Mode of use—Physiological and therapeutic effects—Diseases and conditions treated—Safety of treatment in cases of weak heart—Conditions of successful use—A new departure in therapeutics 1

CHAPTER II.
RHEUMATIC ARTHRITIS.

Mr. Jonathan Hutchinson's opinion—A common and intractable disease—Remarkable success of the Tallerman treatment—Notes of twenty-seven cases from various hospitals and private practice 14

CHAPTER III.
RHEUMATISM, ACUTE AND CHRONIC.

Value of the superheated dry-air method in old-standing cases, with involvement of the joints—Notes of thirty-four cases . 46

CHAPTER IV.
GONORRHŒAL RHEUMATISM.

Rapid recovery reported by Dr. Dejérine, Professor Landouzy and Mr. Alfred Willett, F.R.C.S.—Notes of eight cases . 71

CHAPTER V.

STIFF, PAINFUL AND ANCHYLOSED JOINTS — TUBERCULOUS AND TRAUMATIC INFLAMMATION—SYNOVITIS WITH EFFUSION—STIFFNESS AFTER FRACTURES AND OTHER INJURIES—PERIOSTITIS.

PAGE

Notes of twenty cases - - - - - - 76

CHAPTER VI.

GOUT.

In acute attacks pain and inflammation rapidly reduced—In chronic conditions morbid products eliminated by increased excretion through the kidneys—Observations by Dr. Chrétien and others—Notes of eighteen cases - - - - 88

CHAPTER VII.

SCIATICA—LUMBAGO—LOCAL PARALYSIS—WASTING—WRITER'S CRAMP—CHOREA.

Notes of fourteen cases - - - - - - 104

CHAPTER VIII.

SPRAINS AND OTHER INJURIES.

Frequency of sprains among manual workers and from bicycle accidents—Their troublesome character—Rapid cure—Notes of fourteen cases - - - - - - - 116

CHAPTER IX.

CHRONIC ULCERS.

Encouraging results—Notes of three cases - - - 121

CHAPTER X.

FLAT-FOOT.

Paper by Mr. W. J. Walsham, F.R.C.S. - - - - 123

CHAPTER XI.

REPORT OF THE TALLERMAN FREE INSTITUTE, OPENED FOR THE TREATMENT OF THE NECESSITOUS POOR.

Particulars of seventeen cases - - - - - 129

CHAPTER XII.

MEDICAL REPORTS.

Clinical Lecture at St. Bartholomew's by Mr. Alfred Willett, F.R.C.S.—Paper by Professor James Stewart, of Montreal—The British Medical Association—Report by Professor Landouzy—Report by Dr. Dejérine—Paper by Dr. Édouard Chrétien—Report from the North-West London Hospital—Cases shown at the Harveian Society—Philadelphia Medical Society—Paper by Dr. Knowsley Sibley—Report from Liverpool Workhouse Infirmary—Report from the Livingstone Cottage Hospital—Letter to the *Lancet* from F. Fitzherbert Jay, M.D. - - 135

APPENDIX.

NOTE BY THE INVENTOR - - - - - - 171

THE TALLERMAN TREATMENT.

CHAPTER I.

INTRODUCTORY.

THE Tallerman treatment is a new method of applying heat locally to the cure of disease and the relief of pain. The most important feature of the invention is the use of perfectly *dry air*, which enables a much higher temperature to be applied to the part affected than has hitherto been possible, with correspondingly powerful results. Heat has always been recognised as one of the most valuable therapeutic agents which Nature has placed at our disposal, and various means have been devised for employing it both locally and generally in the service of medicine and surgery. The bath, the fomentation, the hot bottle, and the poultice, are instances familiar to everybody; but, indispensable as these things are, their efficacy is necessarily limited by the comparatively low temperature which is all that can be borne when applied by ordinary means. Hot water becomes painful at about 115° Fahrenheit, and vapour or steam cannot well be borne above 120°; whereas it has been ascertained by experiment that hot air, when dry, can be tolerated up to 300°, and even higher. Attempts have accordingly been made to utilize this fact by so-called hot-air baths of different kinds, of which the best known is the hot chamber (*calidarium*) of the Turkish or Roman bath. In this, of course, the hot air is breathed, but portable baths have been tried in the form of a box, with an aperture for the head, while the air inside is heated by a lamp, thus avoiding the disadvantage of making the patient inhale the superheated atmosphere to which it is desired to subject his body. Similar appliances on a smaller scale have also been used for single limbs. But they have all turned out practical failures. The difficulty has always been to make the high temperature a reality. Either the construction of the apparatus will not permit a high temperature to be obtained at all, or the air is not really dry, and therefore a high temperature cannot be borne when obtained. The air may be dry at first, but as soon as the skin begins to perspire it becomes more or less charged with moisture until the perspiration, which is excessive in the hot-air bath, can no longer evaporate freely, with the result that discomfort at once ensues and the proceeding has to be abandoned. Thus it happens that the hot chamber of the Turkish bath is practically limited to a temperature of about 170°. Even a single limb cannot be exposed to a really high temperature in an ordinary closed chamber for more than a few minutes, for the simple reason that the perspiration resting on the skin has a scalding effect when 200° or less is reached.

Description of the Apparatus.

These difficulties are completely overcome by the Tallerman apparatus, in which a temperature of from 250° to 300° or more can be borne for an indefinite time, not only without discomfort, but with a sensation of pleasure. The secret lies in an ingenious arrangement for keeping the air really dry. The apparatus consists of a copper chamber, varying in size and shape according to the part which it is desired to treat, but generally taking the form of a cylinder. The hand, the forearm, or the whole arm, including the shoulder, may be placed in it; and similarly with the lower extremity—the foot, the knee, and the hip may be treated. The treatment may also be applied to the abdomen and other parts of the trunk. The limb to be treated is passed into the open end of the cylinder through an airtight curtain, which is afterwards secured in such a way as to close the chamber completely. The distal end of the cylinder is furnished with an opening, so contrived as to permit the escape of the evaporated moisture at intervals. In this way the air is kept dry and scalding prevented.

The heat is supplied externally by gas-jets. Oil could be used where gas is not laid on, but the latter is more convenient. The stand on which the cylinder rests is supplied with rows of gas-burners, which can be connected with any ordinary gas supply by a piece of rubber-tubing, and turned up or down according to the degree of heat required. The apparatus can therefore be used in any private house, under the care of the patient's own medical attendant.

The temperature is indicated by a thermometer, the bulb of which is inside the cylinder, at the level of the limb to be treated, while the scale passes outside, and can be read off at any moment by a glance. The cylinder is also furnished with stopcocks, which may be used with or without an air-pump either for drawing off heated vapour or admitting medicated vapour.

Method of Using the Apparatus.

The limb to be treated rests at the bottom of the cylinder on a metal cradle protected by asbestos, which prevents all danger of scorching the skin by contact with the heated metal. The patient may lie in bed or be seated in an armchair during treatment, whichever is considered most comfortable or suitable to the case. The most convenient way of administering the treatment is to heat the cylinder up to 150° Fahrenheit before inserting the limb and then gradually raise the temperature, the process of drying the air being frequently repeated, which enables the patient to bear exposure to 250°, 300°, and even higher, without the least discomfort. This striking result is clearly due to the system of ventilation employed, because each time it is brought into play a considerably higher temperature can be borne without incommoding the patient, until the air again becomes charged with moisture evaporated from the skin. The plan for keeping the air dry is the most distinctive and valuable feature of the Tallerman bath, although in other respects it is a much more complete, reliable, and convenient apparatus than anything else of the kind yet brought forward.*

The temperature found most suitable for treating the majority of cases ranges from 240° to 280°; and the duration of an ordinary sitting is from thirty minutes to one hour, but it may be prolonged to two hours or more without discomfort or ill-effect.

Effects: Physiological and Therapeutic.

In general terms, the effects are those of heat, affecting first the part treated, and then, to some extent, the whole body.

They are thus described by Dr. Knowsley Sibley, senior assistant-physician to the North-West London Hospital, who has given great attention to the investigation of the subject: '*Locally*—(1) The heat produces dilatation of all the cutaneous vessels and free circulation through the parts—it is impossible to say how deeply into the tissues this extends, but from the results it may be judged to be some distance—and at the same time there is a marked stimulation of the nutrition of the cutaneous nerves; (2) there is free perspiration of an acid sweat; and (3) relief from pain, however produced, is almost at once apparent. *Generally*—(1) There is profuse perspiration and dilatation of vessels; (2) increase of the rate of the pulse and force of the heart's action; (3) increase (slight) of the respiratory movements; and (4) an increase in the body temperature,

---

* The late Dr. Bell Hunter, the eminent authority on hydro-therapeutics and medical superintendent of Smedley's hydropathic establishment, Matlock, writing under date March 18, 1894, observed: 'We have our hot-air baths, and with arrangements for local application of the heated air. Your apparatus is a great advance on these in precision of working and extent of range of temperature, however, and I am not unwilling to give it the welcome I can see it deserves.' Mr. Oscar Jennings, writing to the *Lancet*, November 3, 1896, says: 'With ordinary apparatus a local hot-air bath is nothing more than a vapour bath, and all those who are practically acquainted with the question will recognise the value of the Tallerman invention, if it really provides us with a means of keeping the hot air dry and reaching a temperature of 300°.

often of two or three degrees Fahrenheit. The treatment appears to lower the blood-pressure of the body, and in some way to increase the alkalinity of the blood, which enables it to dissolve the uric acid from the tissues and joints, and get rid of this substance through the various excretory organs. This is evidenced by the relief from local pain, and the removal of the frequent uric acid nerve-depression. Hence the treatment is of a tonic nature, and bestows an increased general vitality upon the patient.'*

Dr. Édouard Chrétien, commenting on the cases treated at the Laennec and Salpêtrière hospitals in Paris, states the effects observed in the following terms, which are virtually identical with Dr. Sibley's account:

(1) More or less profuse perspiration, not only of the part treated, but over the whole surface of the body.

(2) Marked reddening of the skin on the part treated, indicating an intense dilatation of the bloodvessels.

(3) Diminution and rapid disappearance, sometimes almost immediate, of pain.

(4) Restoration of movement, where the functional impotence was due only to pain.

(5) More or less marked acceleration of the pulse, caused evidently by the peripheral dilatation of the bloodvessels, which facilitates the action of the heart, and makes it contract more vigorously.

(6) Temporary elevation of the body temperature.†

The increased local circulation, relaxation of the tissues, and free perspiration, are the ordinary effects of the application of heat, and call for no special explanation. But it is to be observed that, owing to the very high temperature to which the part is continuously subjected for a considerable period of time, they are all produced in a much higher degree than by any other existing method. That is, indeed, the distinctive peculiarity of the treatment as a local application, and to it much of the therapeutic effect must be ascribed. The elimination of morbid products is more complete, the improvement of nutrition more rapid,

* The *Lancet*, August 29, 1896.
† *La Presse Médicale*, December 26, 1896.

and the reaction more powerful, in proportion to the temperature employed and the length of time it is maintained. But that is by no means all: the local effects are accompanied by general ones, which are less easy to understand; the whole circulation is quickened and the body temperature raised, with all the results involved in those changes. These phenomena have been repeatedly attested by various observers, and particulars will be found in the case notes; but the point is so interesting that we append here a series of detailed observations made by Dr. Sibley:

TABLE GIVING PARTICULARS OF CASES.

No. of Case.	Date.	Pulse before Bath.	Temperature before Bath (in degrees Fahr.).	Pulse during Bath.	Temperature during Bath (in degrees Fahr.).	Temperature of Bath (in degrees Fahr.).	Part Treated.
1.	Nov. 11, 1895	72	98·0	88	99·4	240	Right hand
	Nov. 14, 1895	76	98·4	90	99·6	240	,, ,,
	Nov. 16, 1895	76	98·2	92	99·8	248	,, ,,
	May 23, 1896	80	98·0	92	99·6	240	Right leg
	May 26, 1896	88	98·6	92	99·8	240	,, ,,
	May 28, 1896	80	98·4	92	100·0	260	,, ,,
	June 1, 1896	80	98·2	96	99·8	250	,, ,,
	June 4, 1896	76	98·2	88	99·0	250	,, ,,
2	April 22, 1896	96	99·0	100	99·6	238	Right hand
	April 24, 1896	88	99·2	96	100·0	240	,, ,,
	April 25, 1896	92	99·0	116	100·0	250	,, ,,
	April 27, 1896	92	99·0	100	100·0	250	,, ,,
	April 28, 1896	96	99·4	104	100·0	240	,, ,,
	April 30, 1896	88	99·0	96	100·0	250	,, ,,
	May 1, 1896	92	98·6	100	99·6	250	,, ,,
	May 5, 1896	84	98·2	92	99·4	250	,, ,,
	May 6, 1896	84	98·4	96	99·2	240	,, ,,
	May 8, 1896	92	98·4	104	99·2	270	,, ,,
	May 11, 1896	92	99·0	104	100·0	260	,, ,,
	May 13, 1896	92	99·0	108	100·4	260	,, ,,
	May 15, 1896	96	98·6	108	99·6	260	,, ,,
	May 18, 1896	92	98·6	100	99·8	270	,, ,,
	May 20, 1896	92	99·0	108	100·0	260	,, ,,
	May 22, 1896	92	99·0	104	100·0	260	,, ,,
	May 26, 1896	80	99·0	92	100·0	260	,, ,,
	May 28, 1896	92	99·0	104	100·0	260	,, ,,
	June 1, 1896	88	99·0	100	100·0	260	,, ,,
	June 3, 1896	88	99·0	100	100·2	230	Left leg
	June 5, 1896	88	99·0	96	100·2	240	,, ,,
	June 8, 1896	92	99·0	104	100·2	240	,, ,,
	June 16, 1896	99	98·4	104	99·4	250	Right arm
	June 22, 1896	92	99·0	108	100·0	240	Left arm

Introductory.

No. of Case.	Date.	Pulse before Bath.	Temperature before Bath (in degrees Fahr.).	Pulse during Bath.	Temperature during Bath (in degrees Fahr.).	Temperature of Bath (in degrees Fahr.).	Part Treated.
3.	June 19, 1896	78	98·4	92	100·0	220	Right arm
	June 22, 1896	78	98·4	92	100·0	200	Left leg
	June 23, 1896	84	98·2	90	100·0	220	,, ,,
	June 25, 1896	84	99·0	100	101·0	230	,, ,,
	June 26, 1896	88	98·0	92	100·8	230	,, ,,
	June 29, 1896	96	99·6	104	101·0	230	,, ,,
	June 30, 1896	92	99·2	100	100·6	240	,, ,,
	July 1, 1896	88	98·0	96	101·0	240	,, ,,
	July 3, 1896	92	99·4	100	101·0	240	,, ,,
	July 10, 1896	84	98·6	102	99·4	220	,, ,,
4.	July 20, 1896	70	98·0	88	99·2	230	Left leg
	July 24, 1896	76	98·4	84	99·0	220	,, ,,
	July 27, 1896	72	98·0	88	99·2	230	,, ,,
	July 29, 1896	76	98·6	88	99·4	200	,, ,,
5.	July 22, 1896	88	98·2	92	99·2	240	Right arm
	July 24, 1896	92	98·6	100	99·6	240	Left arm
	July 27, 1896	88	98·0	100	99·6	260	,, ,,
	July 28, 1896	88	98·6	96	99·6	230	,, ,,
	July 29, 1896	80	97·6	92	99·2	230	,, ,,
6.	July 21, 1896	88	97·2	100	100·0	240	Left arm
	July 22, 1896	72	98·0	88	99·2	240	,, ,,
	July 24, 1896	72	98·2	84	99·0	250	,, ,,
7.	July 20, 1896	72	98·4	100	99·2	235	Right leg
	July 21, 1896	80	99·0	96	100·0	230	,, ,,
8.	July 16, 1896	76	97·4	—	—	240	Right arm
	July 20, 1896	80	98·4	100	99·2	230	,, ,,
	July 22, 1896	80	98·0	96	99·0	260	,, ,,
9.	July 16, 1896	76	98·4	88	99·4	230	Right leg Hip
	July 17, 1896	72	98·2	92	99·0	230	,,
	July 20, 1896	68	98·0	84	99·2	240	
10.	July 28, 1896	100	99·0	112	100·0	220	Left arm
	July 29, 1896	92	98·6	100	99·4	230	,, ,,
11.	July 28, 1896	80	98·0	92	99·2	250	Hips
	July 29, 1896	74	98·2	80	99·2	240	,,
	July 31, 1896	72	98·4	84	99·2	210	,,

Note.—In Cases 4, 5, 10, and 11, and corroborated by isolated observations in several other cases, it was found that the pulse about an hour after treatment was usually slower than before the bath. Occasionally it was of the same rate, but in no instance was it found to be more rapid.*

* The *Lancet*, August 29, 1896.

The quickened circulation may be due in some measure to excitement and, as Dr. Chrétien suggests, to peripheral dilatation of the bloodvessels; but the raised temperature is more remarkable. It will be seen that every operation was invariably accompanied by a heightened temperature in the case of each patient, but that some were more susceptible than others. Speaking generally, the rise is from one to three degrees Fahrenheit. The *Hospital* (August 1, 1896), discussing the point, suggests the following explanation :

'The rise in temperature is an interesting phenomenon, and at first appears in contradiction to physiological teaching. Although not explained by any of the observers who have used this method, it is probably due to imperfectly co-ordinated diaphoresis in parts of the body other than that exposed to the direct heat of the bath— that is to say, the blood as it flows through the heated area is not completely cooled down before it passes on to other areas, where the conditions are different, and where reflex, superficial, vascular dilatation, and consequent diaphoresis, are not correspondingly established.'

However this may be, there can be no doubt that the therapeutic effects are largely due to the increased circulation throughout the body, which produces constitutional changes affecting parts other than those treated, and improving the general health. Numerous instances will be found among the case notes in which the treatment of one limb has effectually relieved others at the same time; and with regard to the general health, the experience of Mrs. L. B. Walford, the eminent novelist, who was treated for pain and failure of strength in the right arm, is worth quoting: 'The treatment for my "poor arm" has operated favourably upon my whole system. I was slightly rheumatic all over. Since undergoing the hot-air cure, I have not felt a twinge of rheumatism. I feel better in every way, refreshed, strengthened' (see p. 114).

This undoubtedly points to constitutional changes, which are probably due in a great measure to the elimination of morbid products both by the skin and the kidneys. With regard to the latter point, some interesting observations were made in the Paris hospitals and recorded by Dr. Chrétien. In a case of long-standing gout the daily elimination of uric acid by the kidneys was found to rise from 57 centigrammes after the fourth bath to 89 centigrammes after the ninth (see pp. 89, 151). Dr. Sibley is inclined to ascribe the relief of pain in a large measure to the same action.

The relief of pain is the most immediate and striking therapeutic effect of the treatment. In some instances it is so remarkable that it must be seen to be believed. Patients racked with pain, which no remedy has been able to remove, are relieved as though by a charm, and the relief is often permanent. At the same time sleep and free movement, which had been banished by suffering, are both restored. In short, the treatment is an extremely powerful anodyne, affecting not only the part treated, but the whole of the body.

Further therapeutic effects are the rapid reduction of inflammation in acute and subacute cases, and the restoration of mobility to stiff joints of old standing where no bony adhesions have formed.

They will be found fully illustrated in the case notes.

Diseases and Affections Treated.

The full uses of the superheated dry-air treatment have not yet been thoroughly explored—experience is constantly bringing new ones to light; but the affections in which its efficacy has been proved are the following:

(1) The arthritic diseases—namely, rheumatism, gout, rheumatic arthritis, gonorrhœal rheumatism, tuberculous inflammation of the joints.

(2) Other affections of the joints—sprains, synovitis, and inflammation due to injury, with their after-effects in the shape of pain, stiffness, and adhesions.

(3) Affections of the nerves—sciatica, tic-douloureux, and other neuralgias, to which may be added peripheral

neuritis, and cases of mal-nutrition from injury to nerves and bloodvessels.

(4) Chronic ulcers.

(5) Flat-foot.

(6) Certain diseases of the skin.

There is reason to believe that in time this list will be considerably extended. For instance, encouraging results have already been obtained in some forms of heart disease, in chorea, and anæmia, and cases of bronchitis and Bright's disease have incidentally been markedly benefited. Various other disorders have been suggested as suitable for treatment, and no doubt many will occur to the medical reader when he grasps the mode of action and learns what has already been achieved. But it is not desired to claim here more than can be vouched for on unimpeachable authority.

ILL EFFECTS.

No serious ill effects have been observed, and very few cases have occurred in which the patients have found the treatment at all trying. Those are probably explained by a careless use of too high a temperature at first, owing to inexperience on the part of the attendant or to the patient's want of intelligence. Of course patients vary in susceptibility, but the temperature is so easily regulated at will that it can be adjusted to suit the most susceptible constitution. So far from complaining of the treatment or being weakened by it, the vast majority of patients find it one of 'luxurious ease,' and are frequently inclined to fall asleep during the operation.

Considering the number of rheumatic patients treated, and the marked effect upon the heart, the important question arises whether there is any danger of untoward results in cases of valvular disease. Happily, all the evidence is the other way. The effect on the heart appears to be distinctly beneficial. In fact, cases of cardiac disease are at the present time being treated at some of the London hospitals; and it is hoped that results will be published before long. Meanwhile, reference may be made to a suggestive and promising trial carried out at St. Bartholomew's.

It had been repeatedly found in treating patients for rheumatism, gout and other affections, that the heart's action and the general circulation were so much improved by the hot-air bath as to encourage its use even when valvular disease was present; and the fact was communicated to Dr. Lauder-Brunton, who had shown in a paper upon the Nauheim treatment, previously published in the *Lancet*, that massage relieved the blood pressure by promoting a flow of blood through the muscles. The same result, it was suggested, could be better and more quickly obtained by the Tallerman treatment, and it was decided to test the method as soon as a suitable case was found. A case was found early in January in the person of a bookseller then lying in the 'Rahere' ward of the St. Bartholomew's Hospital; it was one of aortic regurgitation and obstruction and mitral regurgitation and obstruction, and was described as utterly hopeless, and that the man must die very shortly. Under the circumstances it was considered proper to afford the patient an opportunity of trying whether he could derive any benefit from this new treatment, and on January 14, 1896, the apparatus, which had been in use at the hospital for between three and four years, was taken into the ward, and the patient was treated with Mr. Tallerman's assistance, in the presence of Dr. Lauder-Brunton.

At the commencement the temperature was 97·1° and pulse 94.

After thirty-five minutes the temperature was raised to 97·7°, and the pulse was considerably firmer. The patient had been comfortable throughout. The treatment was continued for a fortnight, during which so great was the improvement that hopes were entertained that some permanent benefit might be derived; at the expiration of that time, however, the patient relapsed into the old condition.

This temporary improvement in a case in which the heart

might be said to have practically reached the last stage of organic disease can only be regarded as most satisfactory, and as indicating that more permanent benefit may be expected from this treatment in cases if taken at a less advanced stage. It also demonstrates the safety with which the Tallerman treatment may be administered in gout, rheumatism, and other cases in which cardiac weakness or valvular disease is present.

FAILURES.

Like every other method of treatment in existence, this has had its failures. Here and there cases, apparently suitable, turn out refractory and derive no benefit. That must always be so. But it is to be observed that some apparent cases of failure are due either to their being quite unsuitable or to an improper use of the apparatus. It requires some intelligence, and considerable care for its effective application, and in certain instances it also requires a fair amount of patience. Persons applying it carelessly and indifferently to the first chance patient they come across are not likely to get a good result. It is necessary to say this because certain institutions, after a perfunctory trial, or none at all, have flippantly thrown the thing aside. A good instance is St. Thomas's Hospital, which returned the apparatus with a curt intimation that they 'had no use for it.' From that very hospital a patient was discharged as 'incurable.' She was suffering from rheumatic arthritis. Nearly every joint in her body was affected, she could not kneel, walk, or raise her arm higher than her chin. She had nothing but a life of hopeless misery and suffering before her, until she sank into the grave worn out by pain or starvation. As it happened, she came under treatment. She has now regained the use of her limbs, she can kneel, walk up and down stairs, and raise her arm above her head; all active mischief in the joints has ceased, pain has disappeared, and her general health has greatly improved. Particulars will be found on page 22. How many like cases are there at the gates of St. Thomas's Hospital? And what is the notion of their responsibility to the public entertained by the gentlemen who could find no use for the appliance which has restored their 'incurable' patient to health and activity?

In conclusion, it is submitted, on the evidence which follows in the succeeding chapters, that the Tallerman treatment is virtually a new departure in therapeutics and a valuable addition to the existing means of dealing with a class of cases in which medicine has often to confess itself peculiarly powerless. In particular, it offers a rapid and effective method of treating those intractable arthritic affections which are too often abandoned as beyond the reach of aid, after every known remedy has been tried and failed. Such cases, if they possess means, are usually sent about to this and that expensive watering-place at home or abroad, at great cost of time and money, in the forlorn hope that they may perhaps derive a little benefit; if they are poor, they pass from one hospital to another, to find their last refuge in the workhouse infirmary. In either case their lot is pitiable in the extreme. The advantages of an expeditious remedy, which can be applied at home with the certainty of relieving their suffering and a good prospect of restoring their activity, need no emphasizing.

CHAPTER II.

RHEUMATIC ARTHRITIS.

FULL details are here given of twenty-seven cases of this disease, which is also called 'rheumatoid arthritis' and 'arthritis deformans,' treated at the following hospitals: North-West London; St. Bartholomew's; Laennec Hospital, Paris; Royal Portsmouth; Union Hospital, Cork; Royal Victoria Hospital, Montreal; St. Mary's; Livingstone Cottage Hospital, Dartford; Children's Hospital, Bristol; at the Institution at 50, Welbeck Street, and by private practitioners.

It is a common disease in this country. As Mr. Jonathan Hutchinson has said: 'England is the home of rheumatic gout; no place beats it excepting Ireland.' It is also a very painful disease, and the results of ordinary treatment are at least as unsatisfactory as the disease is common and painful. 'In a severe case,' to quote Mr. Hutchinson again, ' it must be admitted that the prognosis, under the ordinary conditions of life in England, is a gloomy one.' In point of fact, the prospect is all but hopeless. The malady is of a progressive character, and as one joint is involved after another the patient becomes a hopeless cripple, to die eventually a lingering and painful death, worn out by constant suffering.

In dealing with this intractable disorder, the Tallerman treatment has met with the most brilliant and unequivocal success. Case after case will be found of the severest kind, which had resisted every other remedy and been pronounced incurable, but which has yielded to superheated dry-air treatment. Not only has all pain been removed and the progress of the disease arrested, but a large degree of mobility has been restored even in the most desperate cases, and experience goes to show that the amelioration is of a permanent character. In the *Lancet* of August 29, 1896, Dr. Sibley writes as follows:

'It must be admitted by those who have had much practical experience of severe cases of arthritis deformans such as cases No. 1, 2, and 3, how very hopeless these have generally been considered, and what little chance of permanent good medical treatment offers, that this treatment by hot air at a very high temperature meets a general want. I must add I have never seen results so immediate and satisfactory produced by any other treatment. It is now two years since treatment by the local hot-air bath was commenced in the first case, and yet the patient continues comparatively free from the complaint, and even the deformity of the fingers has greatly disappeared.'

The reader can judge from the case notes which follow how fully this experience of the treatment is borne out by the observation of others.

CASE 1.

Treated at the North-West London Hospital. Under the care of Dr. Knowsley Sibley.

Arthritis deformans; duration four and a half years.—In January, 1894, a single woman, aged 64 years, came under treatment. The patient's father died at the age of 66 years, from bronchitis; a sister, since dead, suffered from rheumatism. The patient had suffered from bronchitis for twenty years. About four years ago she had a good deal of trouble and then influenza, and ever since this time she had suffered from rheumatism, which came first in the ankles, then in the knees, and finally in the hands. This had gradually become worse. There was no cardiac murmur, but there was a good deal of general bronchitic rhonchi. She had been under private treatment, and had been attending University College Hospital for four months, and then came to the North-West London Hospital, at which place she attended regularly from January till August, when the baths were first tried. During this time the following drugs were prescribed in the order given: colchicum, salicylates, iodide of potassium, guaiacum, gentian and rhubarb, salines, strychnia and iron, salicin, antipyrin, exalgin, and lithia, together with various liniments, etc., but nothing seemed to give the patient any relief from the constant pain, or produced any amelioration of the gradually progressing disease; in fact, she was daily becoming worse and more crippled. On August 4, 1894, the following note was made: 'All the finger-joints of both hands are now greatly enlarged, very painful and tender; there is hardly any movement of the fingers, and the patient is quite crippled and unable to use the hands at all, it being impossible to close or separate the fingers. The movement of the wrists is also very limited, and any attempt at flexion causes great pain. The right elbow is fixed at a right angle. Both the knees and shoulders are enlarged and painful, with very distinct grating and considerable disorganization of the joints. The patient is very anæmic, and looks ill and worn out with constant pain. She cannot dress or even feed herself, being unable to get her hands to her mouth.' On August 10, the medicines were stopped and the hot-air treatment commenced. After the first bath the patient said she had less pain than she had had for months; the joints showed some improvement, the fixity not being quite so complete. After the fourth operation the patient, who for a year and a half had not been able to do any work, was again able to use her needle a little. After the fifth bath the right arm could be fully extended without pain, and with a little assistance all the fingers could be flexed on the palms. The left hand, although it had not yet been placed in the cylinder, also showed considerable improvement. On August 24, after the ninth bath the patient reported that she had resumed her former occupation as a dressmaker, and was able to walk up and down stairs with comparative ease and without pain. On September 9, the patient had had twenty baths and considered herself practically cured, but as the apparatus was still at the hospital, she had an occasional bath up to the end of October, 1894. From this time to November, 1895, she continued at her work, and although she had some other ailments (uterine), she suffered no inconvenience from the joint affections. At this time she again complained of some stiffness and pain in the right

CASE 1.—Arthritis deformans, taken in June, 1896 (twenty months after treatment); showing the patient permanently restored to her business of dressmaking.

To face p. 16—I.

CASE 1.—Arthritis deformans, taken in June, 1896 (twenty months after treatment; showing free movement of all extremities and permanency of benefit obtained in 1894.

CASE 1.—Arthritis deformans, taken in June, 1896 (twenty months after treatment); showing free movement of all extremities and permanency of benefit obtained in 1894.

hand, so three more baths were prescribed, and she again rapidly recovered and continued well for some months. *Condition on July* 2, 1896: She has had an occasional bath recently and is practically well. All the movements of the fingers are quite free, and the joints, which were formerly considerably enlarged, are now but very slightly swollen. On July 20, the patient was suffering from headaches, but otherwise she continued to be quite free from pain. She stated that her bronchitis was better than it had been for many years, and the patient had continued uninterruptedly at her work since August, 1894.

CASE 2.

Treated at the North-west London Hospital. Under the care of DR. KNOWSLEY SIBLEY.

Arthritis deformans; duration eight years.—The patient was a married woman aged 69 years, and she came under observation on April 22, 1896. Her father suffered from rheumatism and died at the age of 97 years; her mother died, aged 51 years, from hernia; she also suffered from rheumatism. Four brothers and three sisters had died from old age; of these, one brother and one sister were rheumatic subjects. The patient was the youngest of her family; she had had one daughter, who died in an asylum at the age of 39 years. The patient had suffered from rheumatism and bronchitis on and off for the last eight years, but was especially worse in the winter. There was no cardiac murmur. The hands were especially affected and greatly deformed. All the fingers were enlarged and very painful; for eight months she had been unable to use them for anything, although she had been all this time under medical treatment. The knees and ankles were also stiff and swollen, and she had great difficulty in getting about. On the 22nd the right hand was treated with the hot-air bath, and she was photographed before and after the bath to show the improvement. By May 21 the patient had had fifteen baths and had very considerably improved; she was then able to flex her fingers, and had free movement of the arms and shoulders. The knees and ankles were both much less swollen and painful. Her cough and general nervous condition had much improved. By the 26th she had had seventeen baths, was able to touch the tips of all the fingers with the thumb, and had done some needlework. By July 1 the patient had had twenty-four baths. The hands were quite free and comfortable, and she could do anything with them. There was still a good deal of pain in the left foot and ankle; the varicose veins of this leg were rather prominent; there was also a small and very painful ulcer over the back of the lower part of this leg, which was œdematous. As she was unable to lie up at home, she was taken into the hospital, and remained till the ulcer had healed on July 22. She was then quite free from pain except in the left leg, which still kept swollen, notwithstanding her three weeks' rest. She complained of general weakness from lying in bed, and some strychnia and iron were prescribed, and she left the hospital. On August 6 she had much improved in general strength, and had had no more pain in the hands, the movements of which were quite free and comfortable. She still had pain in the left leg from the varicose veins, which prevented her from getting about.*

CASE 3.

Treated at the North-west London Hospital. Under the care of DR. KNOWSLEY SIBLEY.

Arthritis deformans; duration thirty years.—On June 2, 1896, a woman aged 51 years came under treatment. Her mother died aged 82 years, after being for six years confined to bed with rheumatism. The patient had never been seriously ill, but she had had a mild attack of rheumatic fever when she was 20 years of age, lasting about a fortnight, and she had suffered from rheumatism on and off ever since. The pain had been especially in the knees and hands. Two years ago she had an injury to the left knee from a fall. Since January of this year the pains had become much worse, and had prevented her following her occupation. She had been under the care of several medical men, and, getting no better, she was recommended by her medical attendant to come to the hospital. The hands, shoulders, and knees were especially affected, being much swollen and very painful; the patient suffered a great deal of pain, and was unable to do any work. She could only get about with the greatest difficulty on account of the pain and swelling in the knees. There was no cardiac murmur, but the sounds were rather feeble. She was given salicylate of soda and digitalis for the pain.

* Dr. Sibley, *Lancet*, August 29, 1896.

Case 2.—Arthritis deformans, taken April 22, 1896, before treatment, fingers fixed in above position.

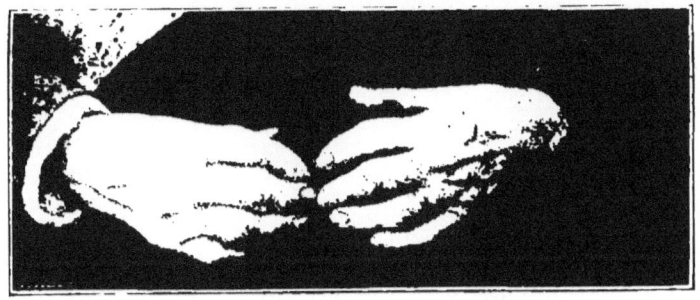

CASE 2.—Taken April 24, 1896, after second operation, showing some separation of fingers.

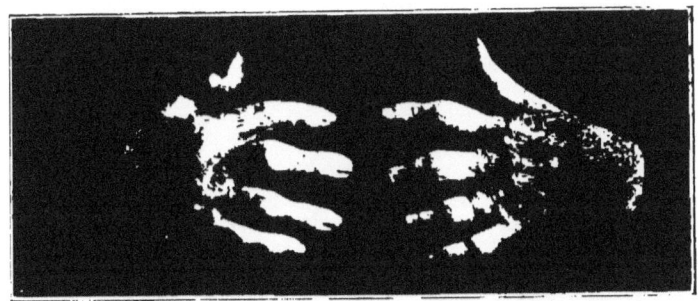

CASE 2.—Taken August 7, 1896, showing result of treatment.

To face p. 18—II.

Until the 17th the patient had been gradually getting worse, and was then unable to raise the right arm or to close her hands, and could only walk, or rather drag herself about, with the greatest difficulty. On the 19th the first bath was given, the right arm being placed in the cylinder. She was photographed before and after the operation. By the 23rd the patient had had three baths, and there was considerable general improvement, the pain and stiffness having quite left the hands and arms. On July 2 there was no pain in the hands or arms, but the legs, especially the left knee, continued weak and painful, keeping her awake at night. She was put on guaiacum and an opium pill at night. On the 10th she had her tenth bath, and with the exception of the knees was comfortable. The fact that the patient lived on the top of the house and had several flights of stairs to mount was very detrimental to the progress of her case, so she was admitted into the hospital on July 23. She felt much better, and only had very occasional pain in the hands. On August 4 the knees had become better, but great difficulty in standing still continued; there was slight pain in the hands first thing in the morning, otherwise these were quite comfortable. The temperature was usually between 99° and 100° F. Strychnia and iron were administered, and the knees were strapped with Scott's dressing.*

Case 4.

Treated at St. Bartholomew's Hospital. Under the care of
Mr. Alfred Willett, F.R.C.S.

Alice C., 25 years of age, was admitted into President Ward on September 29. Her history was one of gradual onset of pains seven years previously, a dull aching by day and night. Six years ago, or one year after the onset of these pains, she had rheumatic fever. She was under treatment at Reading, and subsequently at Bath. On admission, her condition was that of an anæmic girl, with very pained expression; heart's sounds natural. She was almost helpless and bedridden from crippled joints. The affected joints were the right ankle, knee and hip, the left hip and both elbow and wrist joints; all these were stiff and painful. The right hip and knee were flexed at angles respectively of 165° and 90°; both elbows were flexed to

* Dr. Sibley, *Lancet*, August 29, 1896.

about 120°, with about 35° angle of movement. The wrists were completely anchylosed. The changes chiefly affected the fibrous structures of the joints, but about the elbows, especially the left, were bony and cartilaginous defects, with creaking and grating on movement in both. The electrical reactions were normal.

On January 17 the right hip, knee, ankle and right elbow and wrist joints, were all forcibly broken down and moved under an anæsthetic. The right hip and knee were brought straight and fixed upon a Thomas's hip-joint splint. On the 25th, under passive movement there was some improvement. Nitrous oxide gas was again administered for wrenching the right wrist and elbow. At this time the left elbow was the worse, but I thought it better to confine myself to the joints on the right side of the patient's body. She was still being treated by passive movement and given cod-liver oil. On February 20, Tallerman's hot-air bath was used at a temperature of 260° for fifty minutes. There was much less pain afterwards; the range of motion was not increased; the movements, however, were more free through the given range. But on any stretching of the joints, with the view to obtaining increased range of movement, the old pain was felt. The baths were continued from February 20 to March 21. Up to this time there was no very definite improvement beyond the above changes. On April 9 gas was administered, and the left elbow extended and flexed, all the adhesions in it being broken down, and afterwards placed in the bath. On the 19th it was noticed there was general improvement in all her joints. The left elbow, which was the worse, was much less painful and more mobile.

I now show you the patient, who, as I mentioned, was admitted practically bedridden and helpless just at the end of last year. She left the hospital walking on crutches. Any movement was attended with great pain when she came here; on leaving, the joint movements were much freer, almost painless, and the patient was able to do almost everything for herself. These noticeable changes, relief from pain and freer movement, were mainly attributable, I think, to the influence of the hot-air baths. She was under the bath treatment for about eight weeks. On asking her to strike at my hand, we find that the right upper extremity is the stronger, but with both she can hit my hand a fairly vigorous blow; on admission it would have been absolutely impossible for her to have attempted

CASE 3.—Arthritis deformans, taken June 19, 1896, before treatment: showing (i.) highest point hand and arm could be raised; (ii.) inability to flex the fingers; (iii.) expression of pain.

CASE 3.—Arthritis deformans, taken June 19, 1896, after the first treatment, showing full extension of arm and forearm, flexion of hand, absence of pain.

1

anything of the kind. I mentioned the great creaking which occurred in the joints; now within the range of movement the joint has become quite natural. The range of movement in her left elbow-joint is better than it is in the right; the flexion is a little greater, the extension perhaps not quite so good, but here again the movement, although not absolutely smooth, or so smooth as in the right, is accompanied by less marked creaking than when admitted. The wrist-joints were absolutely fixed. She has a moderate amount of pronation and supination; still, if I put any strain upon the joints, it is obvious that it pains her. Her right wrist remains very stiff.

CASE 5.

Treated at the Laennec Hospital, Paris. Under the care of M. OULMONT.

Chronic Deformatory Rheumatism of long standing.— Jean P.; 49 years old. Illness began in 1885. Fingers and toes deformed; anchylosis, with an abnormal position of the large joints; slight movements are extremely painful; the arms hang close to the body, and the knees are bent. For two years P. has been confined to bed, and has been absolutely helpless.

The patient was treated fifteen times by superheated dry air. At the beginning of the treatment the pain almost entirely disappeared, We could see little by little the return of the limited movements of elevation and abduction of the arms, and the flexion and extension of the knees. The fingers also partially recovered the powers of flexion, extension, and resistance. The patient could lift his hands to the back of his head; he could use his hands to cut his bread and meat, to eat and drink, all of which acts had been impossible for him for two years. As for the legs, we cannot say that such results were obtained, but we must attribute that to the existence of several fibrous adhesions; the patient could, however, lift his legs a little from the bed, and his pain had entirely disappeared.

CASE 6.

Treated at the Laennec Hospital, Paris. Under the care of M. OULMONT.

Chronic Deformatory Rheumatism of long standing.— Alfred G.; 43 years old. Illness began in 1888. Considerable deformity of the feet and hands; fibrous anchylosis of the large and small joints, with abnormal positions; excessive pain, which caused the patient to cry out when he tried to move. Eight baths.

Pain disappeared. The patient became able to touch his forehead with his left hand, and to place his right hand behind his head; the hands, which had been closed, were opened, but incompletely. The patient fed himself, which he had been unable to do for a long time. Medium results obtained in the legs.

Case 7.

Treated at the Laennec Hospital, Paris. Under the care of Dr. Oulmont.

Deformatory Rheumatism of the Hands: rapid in development.—Louise B.; 32 years old. The illness began in the middle of the year 1894, in an attack of general rheumatic fever, which attacked more particularly the small joints of the hands and feet. It is impossible to say that it might not be gonorrhœal rheumatism. After the first attack the joints of the hands remained swollen and slightly painful; there was a continual suffering from slight subacute pains, and the hands became gradually deformed.

It was during one of the severe attacks that the patient underwent the treatment. She took seven baths. The pains disappeared almost immediately, so much so that the patient was able to knit. The swelling diminished. The treatment was interrupted by the departure of the patient.

Case 8.

Treated at 50, Welbeck Street. Under the care of Dr. Knowsley Sibley.

Alice Williams; aged 10½ years. The patient had rheumatic fever about fourteen months ago, and was treated in bed for several weeks. She had a serious relapse a few weeks after this, lasting some days. Rheumatoid arthritis then developed itself, attacking in rapid succession all the large joints. The patient was treated by a private practitioner for some time, and then became an in-patient of one of the large general Metropolitan hospitals, where a diagnosis of rheumatoid arthritis was made, and from which, after seven weeks, she was discharged as incurable.

CASE 6.—Before treatment: showing position of hands fixed against chest.

CASE 6.—After treatment.

She came under observation on April 10, 1896, and on examination the following condition was observed:

Hip joints stiff and painful, only allowing very limited movement; there was marked lardosis, with an upward tilting of the pelvis, causing an apparent shortening of the right leg.

Knees.—Considerable thickening round both knees, with slight effusion into the joint. Possible movement is very slight, flexion only being able to be effected to the extent of a few degrees, and at the expense of much pain. The left knee is slightly less affected than the right.

Feet.—Both ankles are much swollen and painful.

The legs are abducted to such an extent that the knees cannot be separated. The patient cannot kneel or walk upstairs.

Wrists are enlarged, stiff, and painful.

Finger joints are characteristically spindle-shaped. The hands cannot be closed.

Shoulders.—The patient cannot raise either hand higher than the chin, and cannot touch the ears.

Neck.—This is very stiff, and interferes with free movement of the head.

The patient can only walk a few steps with evident pain. She is neurotic, restless at night, and occasionally faints.

Treatment was commenced at the Tallerman Institute, in Welbeck Street, on April 7, 1896, and was continued until June 10, 1896, the patient having been treated thirty-four times in all.

After the first operation (on the right arm only) the patient could partially flex the fingers of both hands and raise her arms above her head. After the last bath it was found that the patient was very much improved, being able to stand with the legs extended normally, to kneel, and to cross the legs, and walk up and down stairs, could partially flex the fingers on the palm, and could raise the arms above the head. All inflammatory mischief in the joints appeared to be arrested, as they were now quite painless on forcible movements. The patient had also much improved in her general health.

Condition, July 28, 1897.—Patient has grown and looks very well in herself—in fact, she has nothing to complain of, and is about on her feet all day.

Patient has had no treatment since the baths were discontinued last October (1896), but has continued to improve in every joint. At the present time there appears to be no ordinary movement she cannot execute as well as any child of her age; she can button her boots, pick up pins from the carpet, kneel on the floor, quite dress and undress herself, does needlework very well, and, when standing on the floor, can raise her feet on to an ordinary chair.

Patient has been living at home under hygienic conditions not advantageous to her recovery.

CASE 9.

Treated at the Royal Victoria Hospital, Montreal. Under the care of PROFESSOR JAMES STEWART.

SUBACUTE RHEUMATISM OF FOUR MONTHS' DURATION — MULTIPLE ARTHRITIS, INVOLVING CHIEFLY THE SHOULDER, KNEE AND VERTEBRAL JOINTS — RECEIVED THIRTEEN HIP BATHS — LEFT THE HOSPITAL GREATLY RELIEVED—THE GENERAL NUTRITION MUCH IMPROVED.

A. T. L., aged 42, was admitted to hospital on December 30, 1896, complaining of rheumatism, he giving his history as follows: About the middle of September, 1896, he suffered from headache and feverishness. Towards October 1 pain and swelling set in in the left ankle, and still later in the right ankle and in both knees, giving a clinical picture of acute rheumatism. For about a month he suffered acutely, then the pain disappeared to some extent, but the patient was unable to walk, while with the disappearance of the pain in the joints above mentioned he complained of pain in the spine, and pain and stiffness in the shoulder-joints, so that motion was almost impossible.

Of his personal history, it may be said that he has had repeated attacks of acute rheumatism, and, in addition, has had much mental worry, consequent on business difficulties.

On examination after admission, the patient was seen to be a rather emaciated, poorly-nourished man, with some anæmia, unable to remain in any position for any length of time, and suffering great pain on movement. He was unable to walk without crutches, and then only with great difficulty. The right shoulder joint was very painful on manipulation, and movement in all directions much limited, the same, but to a lesser extent, being the condition of the left shoulder joint. Motion of the vertebræ caused severe pain, chiefly in the lumbar regions. The knees were semi-flexed, and could be with great difficulty extended, but could be fully flexed. There was nothing abnormal about the ankle-joints. In addition to the above, there was a

CASE 8.—Before treatment: showing swollen and crippled joints and general emaciation.

To face p. 24--1.

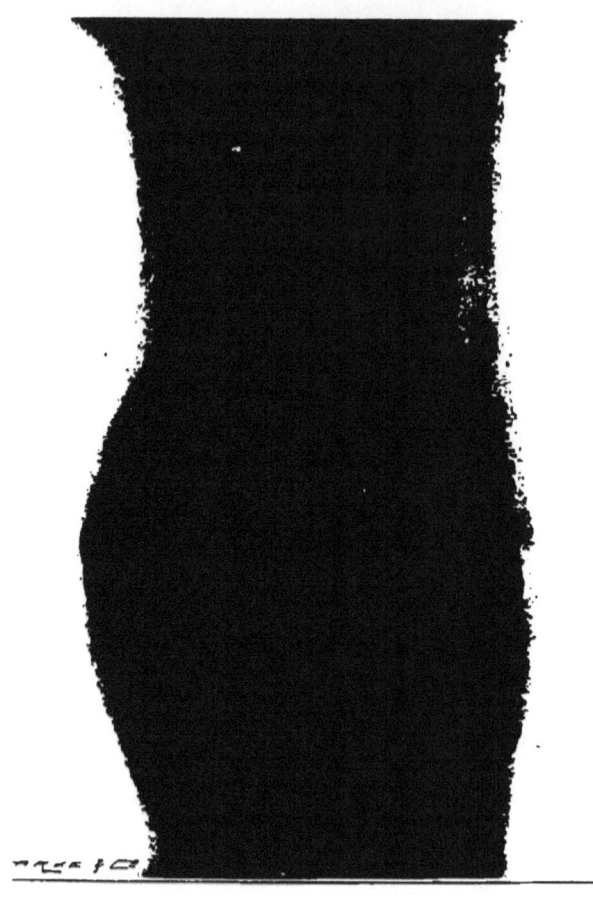

Case 8.—Skiagraph of right knee before treatment.

CASE 8.—Skiagraph of right knee before treatment.

Case 8.—Skiagraph of right hand before treatment.

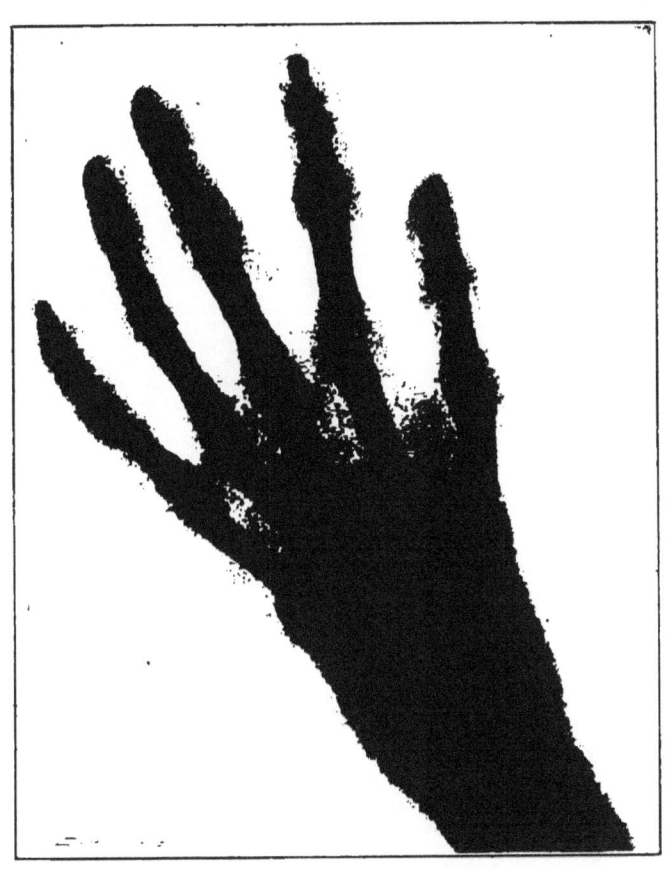

CASE 8.—Skiagraph of left hand before treatment.

To face p. 24—V.

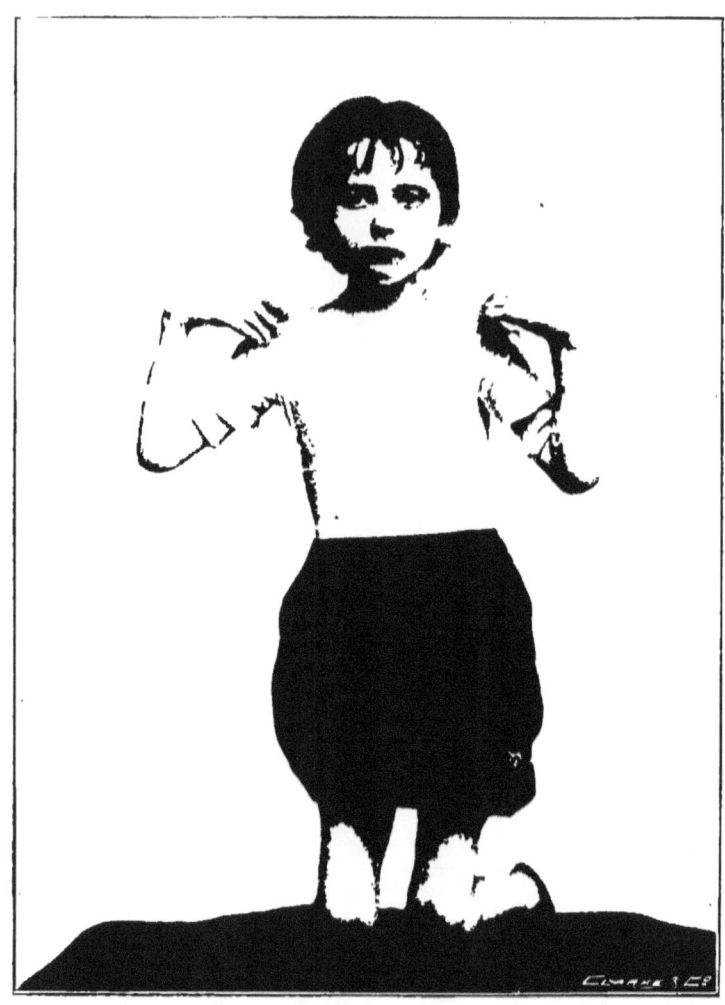

Case 8.—After treatment: showing free movement without pain, reduction of swelling, and great improvement in general health.

CASE 8.—After treatment: showing free movement, etc.

CASE 8.—After treatment; showing free movement, etc.

Case 8.—After treatment: showing easy and natural movement.

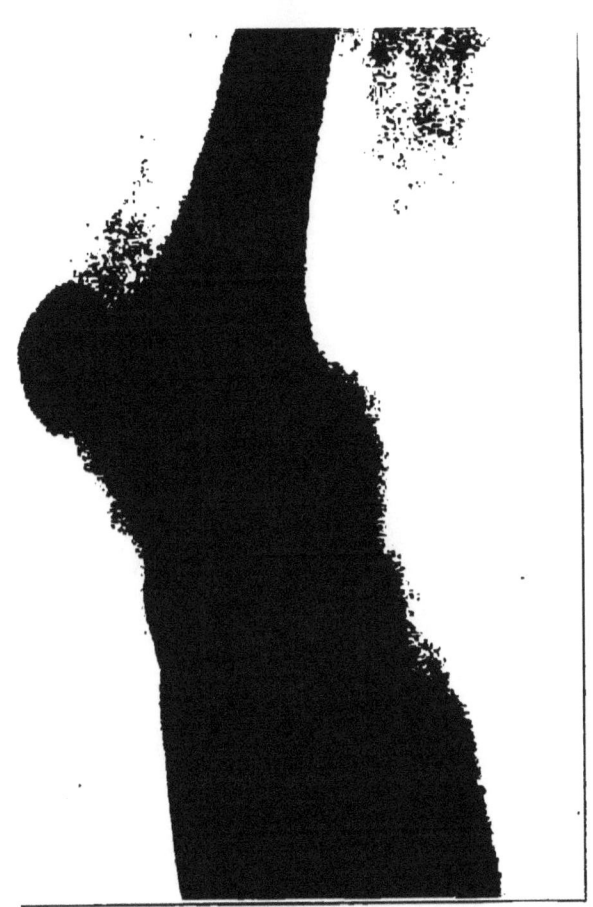

CASE 8.—Skiagraph of right knee after treatment.

faint systolic murmur at the apex, and the pulmonary second sound was accentuated. The other viscera were normal.

The patient was treated by the Tallerman hot-air apparatus, and had in all thirteen baths, with the result that, at the time of discharge, the pain and stiffness had been entirely removed from the knees, and the patient could walk without aid; but he still had slight pain in the back, and the right shoulder was a little stiff, but not painful; the left was free from both pain and stiffness. The appetite and general health were much improved, notwithstanding that the patient had a severe attack of tonsilitis while in hospital.

CASE 10.

Treated at the Royal Victoria Hospital, Montreal. Under the care of PROFESSOR JAMES STEWART.

REPEATED ATTACKS OF SUBACUTE RHEUMATISM—ANÆMIA—EMACIATION—ARTHRITIS OF THE RIGHT KNEE AND SHOULDER—MARKED LIMITATION OF MOVEMENT IN SEVERAL JOINTS—AIR TREATMENT QUICKLY FOLLOWED BY RELIEF TO PAIN—MARKED IMPROVEMENT IN GENERAL NUTRITION.

J. H., aged 26, was admitted to the hospital November 3, 1896, complaining of pain in the back and joints, and difficulty in walking.

The outset of the illness is stated to have been six years previously, at which time he suffered severe pain in the right hip-joint on movement or pressure. Two months later he had acute inflammation in the right knee, and subsequently the ankles and shoulders became involved successively.

After being in bed for nearly a year, the patient began to move about on crutches, and still later was able to get about when using a cane, but not otherwise. The condition remained stationary, apparently, for the next four years, then he began to suffer from pain in the back, with progressive weakness and loss of flesh. In December, 1895, he had acute inflammation in the right knee, and this has persisted to some extent up to the present time.

Save that he had a great deal of domestic worry, his personal history was negative.

On examination he was found to be emaciated and anæmic. There was enlargement of the right knee-joint, due to fluid, and the joint was hot and painful, and, in addition, there was some enlargement of the right shoulder-joint. There was limitation of movement in the left hip, and in the left knee and right hip pseudo-crepitus was easily demonstrated. The patient could not walk without a cane, and when walking there was limitation of movement about the pelvis, the trunk much bent forward, the feet widely apart, and he could not stoop to pick an object off the floor.

The examination of the viscera revealed nothing abnormal. This patient is still under treatment, but after the first application an improvement was noticed. He can now walk about without a cane, has become much more erect, and can pick an object off the floor readily. The effusion into the joints has disappeared, and there is practically no pain either in the joints of the extremities or in the back. His general health has much improved, there being a much better appetite and a rapid gain in weight.

On February 20 the patient left the hospital practically well.

Case 11.

Treated at the North-west London Hospital. Under the care of Mr. Mayo Collier, F.R.C.S.

A. C.; aged 64. Dressmaker. Chronic rheumatoid arthritis of four years' standing; both knees, shoulders, and wrists and all finger-joints affected. Patient had been under hospital treatment for four years without improvement—four months at University College Hospital, and two years at the North-west London Hospital. On examination all the finger-joints were found to be enlarged and painful, and there was great limitation of movement, patient not being able to flex the fingers on the palm. The movement of both wrists was limited and painful; the right elbow was fixed at a right angle, and any attempt at movement caused great pain. Both knees and shoulders were enlarged and painful, and distinct grating could be heard on movement.

This patient was first treated in the hot-air cylinder on August 10. After the first operation it was noticed that the pain was considerably less, and the joints showed marked improvement.

After the fourth operation the patient, who had been unable to work at her business for eighteen months, was again able to use her needle.

After the fifth operation the right arm could be fully extended without pain, and with a little pressure the fingers could be flexed on the palm. The left hand and arm also showed marked improvement, although they had not been treated in the cylinder.

After the eighth operation, with the exception of the forefinger, in which a little stiffness still remained, all the fingers closed readily.

On August 24, after the ninth operation, the patient reported having resumed her former occupation of dressmaker, and her ability to walk up and down stairs without pain.

The average length of each operation was forty minutes.

Note.—This case of chronic rheumatoid arthritis had been under continual medical supervision from its outset, and the usual remedies were applied. Its history shows that so far from yielding to treatment, the disease made rapid strides, so much so that, whereas eighteen months ago the patient was able with some effort to follow her occupation of dressmaker, twelve months later she was incapacitated from even feeding or dressing herself.

The improvement wrought by the local and hot dry-air cylinder was as immediate as it was remarkable. The progress of the disease was arrested, and a curative process was set up at the first operation, which became more manifest with every succeeding one, proving beyond doubt the value of the treatment in cases of this nature.

J. F. SARGEANT, M.R.C.S., M.R.C.P.

CASE 12.

Treated at the North-west London Hospital. Under the care of MR. MAYO COLLIER, F.R.C.S.

G. S.; aged 71. Carpenter. Chronic rheumatoid arthritis of twelve months' standing. Both elbows and all the joints of the fingers were stiff and painful, so much so that it was with the greatest difficulty that he continued his work as carpenter. After five operations the improvement in this case, both as to pain and stiffness, was so marked that he was enabled to go to his work without inconvenience.

Note.—The result of the treatment in this case was quite as satisfactory as in the preceding one.

J. F. SARGEANT, M.R.C.S., M.R.C.P

Case 13.

Treated at St. Peter's Home, Kilburn.

Miss H.; aged 45. For seven years suffering from chronic rheumatic arthritis, which had made her incapable of almost any movement in the hands and arms. After a course of the baths she can now feed herself, and is certainly wonderfully better.

W. D. WATERHOUSE, B.A., LL.D., L.R.C.P.I., etc.

Case 14.

Treated at the Royal Portsmouth Hospital. Under the care of MR. D. WARD COUSINS, F.R.C.S., *President of the Council of the British Medical Association.*

Miss M. suffered from rheumatoid arthritis for some years, being a complete cripple. Knees and elbows stiff, and considerable deformity of the hands, especially the left, which is distorted and useless, the fingers being all flexed and stiff, and any attempt at movement causing severe pain. Unfortunately the patient was unable to undergo full course of treatment; but the results of treatment after two applications of the cylinder were most encouraging. The fingers became more movable and less painful. The little and ring fingers, which had been absolutely stiff and rigid before the treatment, became pliant, and could with perseverance be almost straightened.

T. H. BISHOP, M.B., C.M., House-Surgeon.
H. W. MORLEY, M.R.C.S., L.R.C.P.

Case 15.

Mrs. E. N. G.; aged 31.

History.—Synovitis right knee at 14, left knee at 18, and right wrist at 21, after fall while skating.

Patient's joints, which had been gradually stiffening with much pain during pregnancy, became much worse after her confinement, and left patient entirely crippled and confined to her couch.

Patient states that notwithstanding she availed herself of the advice of several eminent physicians, the disease made rapid strides, and last February found her with both elbows rigidly fixed at an angle of 45°, unable to raise her hand to her face. Jaw very stiff, hardly able to chew; both hip-joints rigid, knees also, with inversion and rotation of the left leg. The thumbs were rigid, and there was no movement at the wrists, and the phalanges bent, so that the hand looked like a bird's claw through the rigidity of the fingers. She was absolutely helpless, and reduced to a skeleton through pain and want of sleep. Most violent palpitation of the heart, with frequent diarrhœa.

The patient, who is evidently of a highly neurotic temperament, then adopted the Salisbury treatment, and at the end of nine months exhibited marked improvement in her joints and condition, the patient digesting well and thriving.

The case was then considered to be one of those which might be treated with advantage by the superheated dry-air cylinder, and a course of twenty baths was prescribed. After which, on examination, it was found patient's knees had extended, and were no longer rigid in one position. The hips are quite flexible, the patient being able to sit on her bed and swing her feet backwards and forwards. She can grasp the head of her bed over her head, do her own hair, hold and raise a cup to her lips. She is bright and cheery, and looks forward with hope to having the power to walk restored to some extent.

The local hot-air baths have quickened her movements and markedly relieved general pain; they have improved her sleep and digestion, and rendered possible the movement of joints which appeared absolutely rigid.

Temperature in this case varied during the application of the bath from $99.8°$, after fifteen minutes, to $100.4°$, temperature at commencement being normal.

Since returning to the country the patient reports herself as much better, and that her muscles are steadily increasing in power and freedom of action. She is able to sit up for meals, which she has not done for three years.

ARCHIBALD KEIGHTLEY, M.D., L.R.C.P.

Case 16.

Miss N. M.; aged 24.

History.—Always weak health; never as strong as the other members of her family, who enjoyed good health. Father not alive—died of consumption.

No trouble with joints till aged 17, except slight stiffness during school life. Nothing noticed until after a dance, when there was great muscular stiffness and some swelling of the joints. Power of movement gradually diminished. Last winter caught cold, which developed weakness in lungs; this was associated with some degree of anæmia. Went to Riviera, where the joints became decidedly worse, and there was no special improvement in the lungs.

Condition in May, 1894.—Lungs, considerable cough (no bacilli), no cavity; some dulness over middle sub-clavicular region on left side. Right ankle stiff, swollen, and painful. Right knee almost rigid, swollen and painful, especially at night. Patella apparently fixed by fibrous anchylosis. This knee was the size of a small cocoanut. Hip is free. Left knee could not be fully extended. The tendons at the inside are rigid and painful, and patella moved very stiffly, with pain on attempting to move it. The swelling was slight, though on each side of the patella there was a 'doughy' feeling on pressure. Left hip free. Shoulders are both free. Both elbows stiff and rigid, very slight swelling. Wrists are enlarged and stiff. Thumbs are stiff and painful. All phalangeal joints are slightly enlarged and flexed. Strictly dieted on meat, very little bread and a little fruit. Oil and a little cinchona.

October 3, 1894.— General health greatly improved. Less pain, better digestion and sleep. Right knee less swollen and painful; rather more pain in left knee; the wrists also were reduced, but the power of supination and pronation had much diminished.

Patient was then advised to take the Tallerman treatment, and a course of twenty baths was taken.

Though of a very neurotic temperament, the patient has received much benefit. Almost all the power of pronation and supination has been restored to the right arm. The right elbow is much less stiff, and moves with fair freedom throughout. The left elbow is less stiff, but pronation and supination only slightly improved. The phalangeal joints are free, though still slightly swollen. The ankles are free. The right knee, though stiff, can be flexed and extended, and the patella is freely movable. The removal of swelling is especially marked in the case of the right knee. The left knee is much more free, very much less painful, and the swelling has almost disappeared. The knees are most markedly improved the moment they are subjected to the hot air. If they are flexed and pain is found on extending, ten minutes' treatment enables them to be almost entirely extended without pain.

Great difficulty was experienced with this patient because the action of the skin on both legs was almost entirely absent. Since this action has been restored, the beneficial effect of the treatment has been much more marked on the knees. There was no difficulty in this respect as regards the body and arms, the joints of which received marked benefit from the first.

The patient, who for three months had been unable to even stand, is now able to walk several times across the room with very little support.

The baths have also had considerable effect in improving the condition of the left lung.

Temperature, always normal at first, varies from 99·4° at commencement to 100·2°.

ARCHIBALD KEIGHTLEY, M.D., L.R.C.P.

CASE 17.

Miss C. C.; aged 54. Suffering from chronic rheumatoid arthritis. All joints affected except the temporo-maxillary, and all those affected rigidly fixed except the left elbow, which is freely movable, and the right elbow, which admits only of limited extension and flexion.

This condition is the outcome of twenty years' suffering, the first onset being an attack of acute rheumatism, from which only a partial recovery was made, succeeded at a short interval by a second attack, from which time more or less pain has always been present with a subacute exacerbation from time to time, and the loss of function of one or more joints till now. *The above condition of utter helplessness has been reached attended always by a constant gnawing pain, increased to acute agony when any movement whatever is attempted.*

A trial was made on November 19 of the Tallerman local and superheated dry-air treatment, the right arm being inserted, whilst the rest of the body was wrapped in blankets. The operation occupied one hour, during which *the pain almost entirely disappeared.* There was a general profuse perspiration, having a pungent characteristic odour. The temperature of the body as taken in the mouth gradually rose from 98·2° to 100°. Pulse rose from 96 to 116.

On the arm being drawn from the bath there was some passive movement in many of the smaller joints of both hands and more freedom of movement in both elbows, the right now admitting of some degree of pronation and supination, whilst that of flexion and extension was increased.

A good night was passed, and next day the patient said that she was 'much lithesomer,' and was able to stand without support, whilst her attendant said that she was lighter to lift and very much easier to nurse.

Some improvement accrued after each bath, which was most noticeable in the feet and legs, the latter, before operations were begun, being rigid and firmly approximated to each other, whilst after the fourth bath they could be separated at the ankles to the extent of ten inches.

The temperature was carefully taken in the mouth every fifteen minutes during the bath, and found to rise uniformly, one degree in the hour, and to fall to normal after the operation had been completed ten minutes.

Unfortunately, it was only possible to treat this patient four times, as Mr. Tallerman, who was good enough to superintend the operations, was prevented by his engagements from prolonging his stay at Southport.

J. G. G. CORKHILL, M.B., L.R.C.P., etc.

CASE 18.

Miss E. R.; aged 21 years. No family history of rheumatism or allied diseases.

1890.—In 1890 first contracted disease in Germany. Returned to England and underwent a course of treatment at Buxton. Got much worse, and was unable to walk. Wintered in Manchester. Adhesions in right elbow and one finger-joint broken; no passive movement was performed after operation.

1891.—Went to Aix-les-Bains; took waters and baths; slight temporary improvement.

Spent summer in Liverpool; treated by splints; got worse; many joints fixed.

Spent the autumn at Grange-on-Sands; got rapidly worse.

In December went to Egypt; slept in damp bed on board ship; arrived in Alexandria in agony with sciatica and pain in various joints.

1892.—Stayed in Alexandria two to three months.

In April went to the sulphur baths of Helouan, near Cairo, for one month; slight improvement. Returned to Cairo; stayed two months; got worse again, and was unable to move or to sleep except in one position in a chair.

In August returned to Europe; took another course at Aix-les-Bains; stayed till end of September; got weaker, however, and was crippled in almost every joint.

In October returned to England. Dr. G. saw her at Charing Cross Hotel, and pronounced her case perfectly hopeless.

1893.—Underwent a course of special diet.

1894.—In January I first saw Miss E. R. at Brighton, and heard that her bodily condition had greatly improved under the special diet, that she had gained flesh, and was comparatively free from pain.

On examination found the following joints affected: Right knee still enlarged, and leg flexed on thigh at an angle of about 130°. Left ankle firmly fixed; also right hip, the thigh being rotated inwards and adducted so that the left knee rested on right thigh. The fingers of both hands all more or less fixed and deformed. Left wrist quite stiff; right wrist partially so. Right elbow quite fixed at an angle of 90°; left elbow allowed slight movement. There was decided tilting of the pelvis, the right brim being nearly four inches below the left, and there was double curvature of the spine.

In February, under an anæsthetic, broke down adhesions in left hip.

This was followed by intense pain, resembling sciatica, for a fortnight. Passive movement was commenced on second day, but accompanied by great pain.

Movements by no means free, and I broke down more adhesions under gas. This caused a renewal of the pain for a few days, but the result was fairly satisfactory, as the legs could now be separated; there was considerable power of flexion and extension, the tilting of the pelvis was much lessened, and the curvature of the spine almost disappeared.

Miss E. R. returned to Brighton, and remained there until October, during which time she greatly improved.

In November, patient commenced the first course of treatment under the Tallerman localized and superheated dry-air treatment. After two preliminary baths to the left arm, I operated on November 26 on the elbow and other joints, and loosened as far as possible the adhesions, paying most attention to those about the elbow-joint.

As soon as the patient recovered consciousness she complained of great pain, but this was quickly relieved when the arm was placed in an apparatus; and after forty minutes' treatment, the arm was taken out and passive movement permitted.

The treatment was continued every day for ten days at temperatures first of 230° F., afterwards 240° and 250°, and finally at 260°.

At no period did the patient experience any pain or uncomfortable sensation from the high temperature.

As the hip-joint had again become rigid—although in a better position than formerly—I prescribed an interval of rest, to ascertain the permanency of the benefit derived before subjecting the patient to further treatment.

1895.—After five months and a half, I found that the condition of the elbow was extremely satisfactory, the patient being able to move her arm freely and without pain, and could lift a weight of 35 lb. I therefore determined to try the effect of the treatment on the hip.

On May 16 and 18 patient had two preliminary baths before operating on the hip, which was now almost rigid again and strongly adducted, and the left pelvic brim was about 2 inches higher than the right.

On May 20, assisted by Drs. Hewitt and Bolus, I broke down (with considerable difficulty) the adhesions until perfectly free movement was capable of being made in every direction.

On recovering from the anæsthetic the patient was placed in the pelvic apparatus, the temperature being gradually raised from 170° to 235°.

The patient complained of very little pain—only of aching—and after treatment in the apparatus for fifty minutes I was able to perform all the movements I had previously done under an anæsthetic.

During the night the patient had very little pain, and on May 21 had another bath, after which passive movements—more rapid than on the previous day—were performed.

The patient had a restless night, but this was most probably due to the catamenia, which appeared on the morning of May 22. The patient could not therefore have a bath, but she was able to move her limb in every direction without pain.

The case is still under treatment.

F. A. BARTON, M.R.C.S.

Case 19.

Mr. P. R. Composer and pianist.

October 9, 1893.—First consulted by patient, and found him suffering from rheumatoid arthritis.

History.—In 1891, the disease attacked the shoulders, elbows and hands; two years later the knees and ankles became involved, the knees being extremely painful at all times and prevented patient sleeping at night.

Prescribed iodide of potassium and usual drugs, and on October 12 consulted Dr. Gower, who approved of the treatment. Continued same until October 25, with little benefit, when, on the advice of Dr. Gower and myself, patient went to Bath for about one month. He derived practically no benefit, and then proceeded to Aix. After undergoing a course of treatment at Aix he returned no better, being unable to walk without extreme fatigue, the walk itself being really a limp attended with considerable pain. On March 20, 1896, the disease having made great progress, the case was considered one in which the Tallerman treatment should be tried.

On Examination, it was found that the left knee was very stiff, greatly enlarged, and painful on any movement; both ankles, shoulders and elbows were slightly swollen and painful; left elbow slightly contracted, left wrist stiff, and painful when flexed.

Right hand.—The fingers were abducted, typical of the disease; and although they could be flexed, the span was limited, owing to contraction of the palm and effusion over the back of the hand.

Left hand.—Fingers were contracted at middle joints, some effusion over third joints. The power in both hands was very limited. Patient had been for some time quite unable to follow his profession of pianist.

After the first operation it was found that the knee was reduced three quarters of an inch in circumference, also that it was less hard, and patient reported that he experienced less pain on movement, especially when walking.

After the second operation patient was able to walk a short distance without any pain, and when that distance was exceeded the pain was less severe. Patient also stated that he felt he had derived more benefit than from the lengthened and varied treatments he had previously undergone.

After the fourth operation patient reported that he

had walked a quarter of a mile without pain. The hands were considerably improved, there being less contraction, less pain, and greater strength.

After the eighth operation the knee exhibited considerable improvement as to shape, and there was a further reduction in size in both knee and ankle. As is usual under this method of treatment, the other joints, although they had not been operated upon, had participated in the benefit: they were smaller and almost free from pain. Patient again expressed his satisfaction with the Tallerman treatment. The hands were now almost normal in shape, with increased strength; the other joints continued to improve.

On May 8 the treatment ceased. Patient walked without pain, his general health had been much benefited, and he was again able to follow his profession of pianist, which he had been unable to do for some time.

Patient was treated twenty times between March 20 and May 8, 1896, at an average temperature of 260° F., reaching on some occasions 300° F.

An active condition of the skin was induced and maintained, which materially aided in the recovery.

The average clinical temperature taken in the mouth (before treatment commenced) was 98° to 98·2°, and it was usually raised during treatment to 100·2°. Pulse beat before treatment commenced about 92 to 96; in about fifteen to twenty minutes it would be found fuller, firmer, and somewhat accelerated, generally to about 112 to 116.

June 18, 1896.—Patient called and reported that the improvement had not stopped with the cessation of treatment, but had continued, and that there undoubtedly was an increasing improvement since May 8, when he had discontinued it. He walked more firmly and without pain, slept well, and had gained in weight; he stated also that his general health had been much improved by the treatment. Patient, who before treatment had been unable to accept professional engagements, and was looked upon as hopelessly incurable, expressed his great desire that the success of the Tallerman treatment in his case might be made known.

     (Signed) L. CROSS, M.D.

Case 20.

Rheumatoid Arthritis, with Chronic Albuminuria; two years.—Mr. A., aged 54, confectioner.

History.— In 1894 patient's kidneys were severely affected, and he was confined to bed until the summer (1895), when he improved, and was able to get about until the return of the cold weather. Rheumatoid arthritis developed during this illness, involving several joints, especially the right knee. The progress of the disease was so marked, and the influence of drugs so unavailing, that the case was considered to be one in which the treatment by the Tallerman Localized Hot-Air Bath might well be tried.

February 5, 1896.—*On examination* it was noted that the right knee was very much swollen (particularly on the inner side), slightly contracted, and extremely painful; patient walked with great difficulty; has at times considerable effusion over the ankles and feet. The right hand was almost powerless, some of the joints were enlarged, and he was not able to flex the fingers to the palm, nor raise a cup to his lips; left hand not so powerless, and could close it; right elbow contracted with 'cracks' on movement. There was chronic albuminuria.

After the first operation the right knee was extremely painful when moved before it was placed in the cylinder, but afterwards I was able to move it freely without causing any pain up to a certain point, and when moved beyond that the pain was only slight. I thought the improvement in the joint was most marked.

After the second operation joint much improved. Patient reported that he could move limb better in bed, and slept better.

After the fourth operation it was noted that the knee was more flexible, and had a wider range of movement; the right hand was much steadier, and patient was able to lift a glass to his lips easily.

After the fifth operation there was some amount of improvement in the condition of the kidneys.

Patient has had an attack of gout over top of right foot, the pain of which was greatly relieved by the bath. This acute attack appeared to have been caused by the elimination of morbid products from the knee; these products, when dissolved out by the hot-air bath, would be eliminated through the skin and the kidneys, and if these organs cannot act freely and rapidly, the uric acid must settle elsewhere.

The general improvement in the affected joints continued; and after the tenth operation it was noted that the right knee had decreased in size one and three-quarter inches; there was great mobility and wider range of movement, with considerably less pain. The hands were more flexible and stronger, and patient's general health was greatly improved; but he was still unable to walk by himself. Skin acted well throughout, and there was copious perspiration.

The patient was treated ten times, the last occasion on March 20, each operation averaging forty-five minutes. Average temperature attained 240° F. It will be noted the body temperature was raised on each occasion, and the pulse beat was accelerated.

Body Temperature.		Pulse Beat.	
Before Treatment.	After Treatment.	Before Treatment.	After Treatment.
99·0° F.	99·8° F.	100	108
98·0	99·2	84	92
98·2	99·6	88	100
99·0	100·2	96	100
100·0	100·6	80	100
99·0	100·0	100	108
98·2	99·6	100	108
98·2	99·4	88	96
98·0	99·0	100	106
98·4	99·2	88	92

Remarks.

The most notable circumstance in connection with Mr. A.'s case has been the marked and rapid diminution of size of the affected joint under the hot-air treatment.

The improvement in the general health has been noticeable also, but the subsidence of the disease implicating the kidneys has not continued. There is still albumin in the urine, but it is not so much as before treatment.

After the tenth bath Mr. A. was put on an alkaline alterative mixture, containing potassium iodide and potassium bicarbonate. This seemed to clear out of the system the morbid products which had been thrown out of the knee by the influence of the hot-air cylinder.

Mr. A. was treated by the Tallerman Hot-Air Bath for the tenth and last occasion on March 20. The improve ment had not alone been maintained, but had steadily increased, and he now, and for weeks past, has been able to walk up and down stairs, to dress himself, and to see after his business affairs, things which he had not been able to do for many months.

H. V. KNAGGS, M.D.

CASE 21.

K. B.; 43. Unmarried.

I have attended this lady for occasional attacks of bronchitis during the last two years. At my first visit I was much shocked at her miserable condition; her head was fixed almost to the left side of the chest—quite fixed, so that she could neither move it nor bear it to be raised at all; her elbows were fixed firmly at right angles; her hands, resting on her abdomen, were painful, joints enlarged; fingers were of various abnormal shapes, some of the joints dislocated, the ring-finger bent backwards from the first metacarpal joint; the knees were swollen, and very painful on the slightest movement—the poor creature was, in short, rolled up almost like a ball. She was a very intelligent person, and conversed with me quite cheerfully.

The rheumatism attacked her more than twenty years ago, but more severely in Jubilee year (1887). With the exception of a slight amelioration in 1890, she has got steadily worse. She has not been able to bear her feet on the floor since 1890. This was a case where medicines of all kinds had been tried, change to various places—Buxton, etc.—without benefit; an incurable person, generally suffering pain.

On November 16 last I managed with much difficulty to place one of the legs in the hot-air bath. She could not bear more than 220°. Next day she told me that she had been more free from pain, and decidedly easier than she had been for a long time. I gave her a bath daily till the 21st. After the fourth bath she raised her left hand to her head; she could raise her head and look up to the ceiling. This she had not been able to do for years. *Three* years ago her father carried her into the garden to look at a beautiful star, but it was quite impossible for her to see it. After fifteen baths—the highest temperature that could be borne was 240°—she could raise both hands above her head, reach things she wanted from the table, and open the hands, which previously had been so contracted that it was scarcely possible to see the palms. Also she could now lie nearly at full length on the sofa, and with the help of two persons she could walk across the room, and she could stand with one person supporting her on one side; she can turn herself in bed as she wishes; for years past she had to awake her sister several times every night to move her.

All rheumatic pains quite disappeared after the first few baths; and when I saw her on the 26th inst. there had been no return, and she told me she was daily improving in her physical powers.

(Signed)   W. J. HODGSON.

CASE 22.

Treated at the Union Hospital, Cork. Under the care of DR. W. ASHLEY CUMMINS.

K. M.; aged 19. Nurse.

Chronic rheumatoid arthritis ten months. Admitted to Union Hospital October 28, 1895, and having been previously under treatment at the South Infirmary, when she took her discharge, her case being considered incurable.

Both elbows, wrists, and most of the joints of the fingers were affected, as also the left knee and both ankles.

Patient has been under hospital treatment for six months.

On examination the affected finger-joints were found to be enlarged and painful, and the movement was greatly limited, so that the patient could not flex the fingers on the palm. Left knee was found enlarged and painful, and the movement of flexion limited. Right knee the patient describes as having been enlarged and painful about five months ago, but the swelling has completely subsided under hospital treatment, she is now free from pain, and can freely flex and extend the joint. Both ankles are swollen, and the patient cannot stand without aid.

This patient was first treated in the Tallerman superheated dry-air apparatus on December 3, her right forearm being placed in the hot-air cylinder for forty minutes; during that time her temperature rose half a degree, and she perspired freely; when the forearm was taken out it was covered with perspiration, and the skin presented a mottled bright-red appearance.

The average temperature of the hot-air bath during the operation was 260°.

Result.—After the first sitting, it was noted that the

pain was very much less, and that the affected joints showed signs of marked improvement, the various movements being considerably freer.

After the second operation—duration seventy minutes—when the left forearm was submitted to treatment in the hot-air cylinder, I was able, with a little force and without causing much pain, to completely extend the forearm, which before had been fixed at an angle of 130°. She can now move the affected joints much more freely.

The patient was comfortable throughout both operations; the second was carried on over one hour, when the patient, hearing that it was to be the last, owing to Mr. Tallerman's departure, expressed the desire that it might be continued.

It was noted that the joints other than the one submitted to treatment showed the same general signs of improvement. After the second bath I found the patient could stand without aid.

(Signed) T. J. MURPHY, M.B.

CASE 23.

Treated at the Livingstone Cottage Hospital, Dartford. Under the care of T. F. CLARKE, M.D.

E. W.; aged 58. The disease commenced eighteen years ago, the left hip being first affected; pains in most of the other joints quickly followed, and were gradually accompanied by swelling and stiffness. The joints became enlarged in the following order: Wrists and hands, twelve years ago; left knee and both ankles, ten years ago; right knee, six years ago; both elbows, three to four years ago; and both hips, two years ago.

On examination patient unable to stand, there being no power in either thigh; both knees enlarged, especially the right; both legs slightly flexed, inability to straighten them or to flex them beyond a few degrees; ankle-joints enlarged and give pain on movement; hip-joints slightly swollen and very painful on movement; wrists and metacarpo-phalangeal joints much swollen and very stiff; fingers slightly flexed and very stiff; elbows slightly swollen, but movement free both here and at the shoulders; movement of lower jaw now and then causes crackling.

This examination was made on March 18, 1896, and the patient's right leg was then placed in the hot-air cylinder, that leg being the most painful and stiff.

Temperature of bath at commencement ...	150°
Patient's temperature	97°
,, pulse	96

After fifteen minutes.

Temperature of bath	180
Patient's temperature	98·6°
,, pulse	105

At the end of thirty minutes.

Temperature of bath	200°
Patient's temperature	98·8°
,, pulse	112

At the end of forty-five minutes.

Temperature of bath	210°
Patient's temperature	100°
,, pulse	116

The patient's leg was taken out at the expiration of forty-five minutes. During the latter part of the time she perspired fairly well, but not profusely; the pulse, though quickened, was quite regular, and was fuller and firmer than at first.

On again examining the patient the right leg could now be all but straightened without pain, there was much freer flexion of the knee-joint, and the hip-joint could also be moved to a greater extent and with but little pain. The ankle-joint now caused no pain on movement. The left leg was also much improved in all its movements, and a similar improvement was noticeable at the wrists and finger-joints.

The patient could also with a little assistance bear more weight on her feet.

Further applications of the bath were made on March 20, 23, 26, and April 5, 9 and 13. It was noticed that with each application it was necessary to lower the temperature of the bath; the average duration of each bath was about forty minutes. The limb inserted never perspired freely.

Since this treatment with the hot-air bath was adopted the patient has been able to move all her joints much more freely, and the swelling of the knees, wrists and hands has subsided to a marked extent. She is also able to stand, though not to walk without assistance.

I would observe that this has been a very extreme case

with which to test the value of this form of treatment, but so far the results have been most promising; and although a cure could scarcely be expected, one might reasonably rely on still greater improvement following further applications of the bath.

September 5, 1896.—Since the baths were discontinued the patient has undergone a course of massage. This has not effected any marked improvement, but at all events has prevented her going back in any way.

I am firmly of opinion, and this is backed up by the patient's own feelings, that further applications of the bath would have brought about considerable improvement in all the symptoms.

Case 24.

Treated at the Children's Hospital, St. Michael's Hill, Bristol, at a Demonstration.

M. A. S.; 64. Well-marked polyarticular rheumatoid arthritis of twenty-five years' standing. Movement limited in all the joints of left arm and hands. Can just raise hand up to head. Right arm, cannot move elbow-joint, cannot close hand. Has great pain in the joints, especially at night, and cannot sleep in one position for any length of time.

First bath, May 12: temperature, 220°; duration, forty minutes.

Patient's temperature: before bath, 98°; after, 100°.
Patient's pulse: before bath, 80; after, 120.

May 13.—Said she had had a better night than for many years. The bath had completely relieved her pain; she could move her hands better, and her feet were also easier.

Second bath: temperature, 250°; duration, thirty minutes.

Patient's temperature: before bath, 98°; after 100°.
Patient's pulse: before bath, 70; after, 82.

After the operation she could close her hand much better, and it was perfectly easy to apply passive movement without giving pain.

E. C. WILLIAMS, M.B. Cantab.

Case 25.

Treated at the same Hospital as the Preceding.

S. S.; 47. Well-marked rheumatoid arthritis of both hands, elbows and shoulders. Could not extend her elbows. Had a good deal of pain.

One bath: temperature, 240°; duration, thirty-five minutes.

Patient's temperature: before, 99·6°; after, 100·4°.
Patient's pulse: before, 102; after, 128.

She said next day the pains were quite abolished by the bath; she was able to sleep, and could move her fingers better. In both the above cases the patients were most emphatic in their statements as to the abolition of pain after the bath. They also said that the other joints were rendered more supple.

E. C. WILLIAMS, M.B. Cantab.

CASE 26.

Treated at St. Mary's Hospital, London, W. Under the care of MR. HERBERT W. PAGE, F.R.C.S.

W. H.; employé G.W.R. The patient was a man who had rheumatism of many joints after exposure to cold; one knee was left swollen, painful, and crippled, and the patient was confined to his bed.

On November 14, 1894, the patient was first treated in the hot-air bath for forty-five minutes at 220° F.

On the 17th he stated that he was 'decidedly better,' and he was discharged after further treatment on several occasions; he expressed himself as having been distinctly benefited by the treatment.

CASE 27.

Treated at the Hospital for the Ruptured and Crippled, New York. Under the care of V. P. GIBNEY, M.D.

Chronic Rheumatoid Arthritis, already improving under Massage and Traction.—A lady, 45 years of age, came under my care May 18, 1896. It was in 1890 that her disease began, and the knees were stiff and painful almost from the beginning. She had patronized several of the baths in the United States, with a slight amount of relief, during the earlier years of her disease. It is only fair to state that from July 1 to November 27, 1896, a vast deal had been accomplished in the way of correcting her deformity at the knees, and in getting some use of her right elbow and of both hands. In addition to massage and traction she had taken large doses of potassium iodide. From July 29 to October 28 she was in solid plaster of Paris bandages, with knees extended to about 155°. When we began treating her in the Tallerman apparatus her range of motion was from 130° in flexion to 170° in extension. She had been able for one or two months to use her hands in feeding herself, after a fashion, and even in adjusting her hair; but the thumbs and fingers were so distorted that she was unable to use the third, fourth and fifth fingers at all, but with the index-finger and the thumb she managed to grasp objects. Heretofore, whenever any attempt had been made to move the hands or fingers, rather violent reaction had set in, and it had been necessary to discontinue. She was subjected to this treatment on May 27. Her temperature was raised 0·2°, her pulse only 5 beats. After this bath she had gained very little, if any, in motion. At present writing she has had four baths, of a temperature ranging from 240° to 260° F. The bodily temperature has been raised less than a degree. She has expressed herself as being very comfortable. Profuse diaphoresis occurs, and for the last two mornings I have employed passive motion of the thumb and fingers without producing any reaction and very little pain. She winces when I break up little adhesions, but tells me that the pain soon passes off, and not only her knees are becoming more flexible, but her hands as well. There has been very little, if any, gain in the extensibility of the legs. She will probably need breaking up of adhesions under an anæsthetic.

CHAPTER III.

RHEUMATISM.

In this chapter thirty-four cases of acute, subacute and chronic rheumatism are given from various sources. Most of them are of the chronic type, with involvement of the joints. These have naturally come under treatment more frequently than the acute forms, because they are regarded as test cases, in which the ordinary remedies have been tried and failed. On that account they particularly illustrate the value of the superheated dry-air method, which has been resorted to as a last resource, and has succeeded where everything else had failed. But there is good reason to believe that if it were tried more often in the earlier and in acuter stages, patients might be saved from the chronic condition altogether. A good many of the cases are quite indistinguishable from rheumatic arthritis, and might have been included under that head; but it has been thought better to classify them according to the nomenclature adopted by the medical gentlemen who have treated them, and are responsible for the case-notes.

Cases 1-7.
Treated at the North-West London Hospital, and at the Tallerman Treatment Institute, 50, Welbeck Street. Under the care of Dr. Knowsley Sibley.

Case 1.

The patient was a girl, aged 26, single. Her mother suffered for fifteen years from rheumatism, and her father died, aged 56, from kidney and heart troubles; her maternal great-grandfather also suffered from rheumatism.

History.—Always well until about four years ago. She suffered from anæmia as a girl. The rheumatic affection commenced gradually in the right little finger, then left big toe, right knee, then the other fingers, and then hands and elbows in succession.

In August, 1894, she went to Bath, and took the baths at the Mineral Water Hospital, and was there seven months; she had baths and salicylate treatment. She became worse, the right knee became contracted and fixed, so the baths were stopped; she had a series of colds, and the rheumatism was worse. She returned to her home in Wales in April, 1895, then in many respects better; but shortly afterwards she again relapsed.

By November, 1895, she had become much worse, and was unable to open her mouth on account of rheumatism in the jaw.

In January, 1896, she went to Brecon; while there was given colchicum, salicylates, iodide of potassium, and tonics.

In September, 1896, she was sent to London for treatment.

State on admission, September 30, 1896: Patient had used crutches for two years; she was unable to walk up or down stairs, to wash her neck, to put her jacket on or off, or to do her hair; she fed herself with a large spoon with difficulty, as she could not get her hands to her mouth; she could only see the back of her hands, and was quite unable to rotate the elbows. Pulse 72, regular; no cardiac bruit.

Hands.—The middle phalangeal joints of both were considerably thickened, the right little finger was deformed at the terminal phalanx, wrists were thickened, and the hands deflected outwards.

Elbows.—Both were almost fixed at right angles, and completely pronated; there was little or no movement of either flexion, extension or rotation.

Shoulders.—Considerable grating in both; movements fairly free.

Right Knee.—Anchylosed nearly at a right angle; absolutely no movement; muscles of this leg and also of the thigh much wasted; the tip of the toes could just be placed on the ground; the limb was powerless—patient could not even raise it off the bed without assistance—and there was considerable thickening around the knee-joint; but there was little or no effusion. The tendons at the back of knee were very rigid, tense and fixed. Patient was unable to place this foot on the ground without the gutta-percha splint round the knee-joint, and even then was unable to bear any weight on this limb. The knee was very painful.

Left Foot.—Œdema on dorsum. Pain on movement.

The first hot-air bath was given on October 1, the right arm being placed in the cylinder.

On October 3, after the second bath, it was possible to rotate the left elbow, so as to supinate the palm of the hand. After the third bath the patient was able to see the palm of her hand, which she had not been able to do for two years, and also to touch her forehead with it.

On October 7, after the sixth operation, patient was able to do her front hair, and there was now some more movement of flexion and extension in the elbows.

On October 12, after the tenth bath, the patient was able to walk a few steps without her crutches; there was distinctly some movement in the right knee-joint, and she was able to take a few steps upstairs.

On October 13, the right leg was placed in the cylinder, and after treatment there was increased movement in the knee-joint.

On October 14, the patient walked out of doors for an hour, and on returning, walked upstairs with the aid of one crutch.

By October 21 patient had had sixteen operations, extending over twenty-one days. The movements of the elbows and wrists were now sufficient to permit her washing and dressing herself, including doing her hair; she was also able to feed herself with ease. All through the treatment she had been practically free from pain, even in the knee-joint, after one had forcibly broken down some adhesions. These active movements had not been accompanied by any effusion into the joint. Her general health had also greatly improved. She has been taking a teaspoonful of the syrup of iodide of iron three times a day, and some natural saline water as a mild aperient when necessary.

On October 22 she had the eighteenth operation, and was shown before the North-West London Clinical Society. She could then put the right foot more firmly on the ground.

By November 4 patient could walk round the room without her crutches, with the help of a stick.

On November 6, after the thirtieth bath, she was shown before the Harveian Society of London, and the movement which was then to be seen in the knee-joint was demonstrated (*Lancet*, November 21, 1896).

CASE 1.—Before treatment: showing fixed position of elbows and inability to flex same.

CASE 1.—Before treatment: showing nearest approach of hands to mouth when feeding herself.

CASE 1.—Before treatment : showing fixed position of leg.

Case 1.—After treatment.

To   p. 48—IV.

Case 1.—After treatment.

On November 10, it being thought desirable to hasten the increase of movement in the right elbow, patient was given a little gas and oxygen, and the elbow forcibly flexed and extended, and the same, to a less extent, was done to the right knee. As soon as she came round the right arm was put into the apparatus. She had little or no pain afterwards, and passed a very good night. The next day there was no effusion into either of the joints which had been moved, and no pain about them. The increased movement was with difficulty maintained on account of the wasting of the muscles, especially of the biceps, through disuse.

On November 20 patient's condition had steadily improved. There had been no rise of temperature or effusion into the joints since the movement under the anæsthetic. She left the home for a fortnight's change, and returned on December 14, when she was able to walk up and down stairs without her crutches or stick.

On January 4, 1897, after the fifty-fourth operation, she walked out without her crutches, only using a stick.

On January 15 patient left for a convalescent home in the Isle of Wight, where she continued to make steady, uninterrupted progress. She remained till the middle of April, by which time she could walk about without a stick for four or five hours a day, and she had had no return of rheumatism in any form.

The patient was also able to dress, undress, and, in fact, do everything for herself, and no longer considered herself a cripple or invalid.*

Case 2.

Subacute Articular Rheumatism; Aortic Regurgitation; Psoriasis.—On September 10, 1894, there was seen a married woman, aged 19 years, with two children. The patient's father was living, but suffered from chronic rheumatism and bronchitis. No other family history of importance was elicited. She had had psoriasis on the knees and elbows since 7 years of age, but had never been treated for it. She had rheumatic fever at 15 years of age, and again at 18½ years, complicated with heart disease. She now had a subacute attack of rheumatism, with much pain, especially in the right shoulder and the left knee and leg. She looked ill, and had an aortic diastolic murmur; the heart was not much enlarged. Salicylate of soda was prescribed and continued to October 4. Not making much progress, and the patient

° *Medical Times and Gazette*, May 22, 1897.

still suffering much pain and inconvenience, the hot-air treatment was tried, the salicylates being omitted. On October 4, the right shoulder was placed in the cylinder, and the next time the left leg. On October 16, the pain had nearly all gone, and she was able to run to catch the tram, a thing she had been unable to attempt for weeks. On the 22nd all pain had gone; the psoriasis appeared to be slightly more extensive. On the 25th she had had six baths at a temperature of about 220° F., the duration of each being forty minutes. The heat brought out an irritable eruption, which, however, soon subsided. On November 11, she had been quite free from rheumatism since the previous date. After this she was for a time under the care of an obstetric physician for some uterine displacement. On November 7, 1895, there had been no more rheumatism. The patient was rather anæmic; the condition of the heart and the psoriasis were much the same. The pulse was 90. Strychnia and iron were prescribed. On July 23, 1896, the patient said she had her third child in May; she was very ill during the greater part of the pregnancy. She was now very pale and worried, the infant being ill. The psoriasis was possibly not quite so extensive as it was formerly. The pulse was 88. The condition of the heart and the aortic murmur had not changed. She had much dyspnœa on quick movement. No more rheumatism had occurred. She was given strychnia and iron.

CASE 3.

Chronic Rheumatism; duration eight years.—The patient, an unmarried woman, aged 61 years, came under notice on October 29, 1894. She had had two attacks of rheumatic fever some years previously. For the last eight years the fingers, especially of the right hand, had gradually become stiff. She continued at her work till a recent date, when the pain and stiffness appeared in the right shoulder, and she was unable to raise her arm, and so was discharged. The right shoulder was very painful and more or less fixed; the right leg was also swollen and painful. There were considerable deformity of the hands and enlargement of the middle phalangeal joints and bones. She was given an alkaline gentian mixture. On the 30th the right hand was placed in the hot-air apparatus. On November 11 the patient had had two baths, the right shoulder was much freer, and there was less pain. She did not come to the hospital again, and I have been unable to find out what became of her.

Case 4.

Chronic Rheumatism; duration ten years.—The patient, who came under treatment on June 6, 1896, was an unmarried woman, aged 59 years. Her grandfather on her mother's side died aged 82 years; her father died aged 40 years, from a chill; her mother died aged 76 years, and suffered slightly from rheumatism; two sisters and two brothers died from phthisis at the ages of 9, 15, 17, and 24 years respectively, and another sister died from cancer of the breast aged 36 years. The patient had congestion of the lungs when 20 years of age, and had suffered slightly from bronchitis ever since. She had a mild attack of rheumatic fever when aged 38 years, which was followed by some left hemiplegia, from which she soon recovered. She had rheumatic fever again when 50 years of age, and another severe attack when aged 52 years. At that time she was in bed for five months, and the heart was said to have become affected. She had been a constant victim to rheumatism ever since that time, the pains being especially severe in the legs and feet. The patient had been obliged to walk with a stick for many years. There was no cardiac murmur, but the first sound of the heart was not clear. An alkaline mixture was prescribed. On July 20, not getting any better, she was ordered a superheated dry-air bath. After the first bath all the pain had gone; she was rather tired the following day. On the 23rd she was much stronger, the movements were freer, and she slept better. On the 24th there was a slight return of the pain in the left knee and toes, and a second bath was administered. On the 27th she had had some pain all over her, but this became better. After the third bath she could walk much better, and could put her foot flat on the ground, which she had been unable to do for many years on account of the contraction of the toes. On the 29th there was some return of the pain, which was relieved by another bath. On the 30th the patient reported herself well.

Case 5.

Subacute Rheumatism; Mitral Regurgitation. — On July 9, 1896, a married woman, aged 32 years, was seen. The patient's father suffered from rheumatism for many years. She was the youngest of six, and the only one who was rheumatic. She had rheumatic fever when 16 years old, and was then in St. Bartholomew's Hospital for about sixteen weeks. After this she was free from rheumatism till two years ago, when she had two attacks of inflammation of the lungs, followed by a mild attack of rheumatic fever, with which she was laid up for three weeks. After this she went to Buxton. She felt better on her return, and kept well till July 6 of this year, when she woke up in the morning with pain in the feet and legs, and then it extended to the arms and shoulders, and she became quite crippled. The patient was anæmic, the pulse was 120, and there was a faint systolic apex murmur. Salicylate of soda and digitalis were prescribed. She continued under treatment, but was not much better on July 20; there was still a good deal of pain. The salicylates were stopped and strychnia and iron prescribed. On July 21 the first bath was given. She could not raise the left arm or close the hands, she had had but little sleep at night for more than a week, and she was very depressed in herself. After the first bath the arm could be fully extended over the head and the hands clenched, and all pain and feeling of depression were gone. On the 23rd she was able to walk home after the first bath, a thing she had not been able to do since the commencement of the attack. She had now had two baths, was quite free from pain, and also felt much better in herself. On the 24th there was some return of pain in the left shoulder; she attributed it to having slept in a draught and having eaten meat the previous day for the first time. After another bath the pain again was all gone. On the 27th she had a return of pain generally. Salicylates were again prescribed. On the 30th the pain had all gone except a little in the left shoulder. Strychnia and iron were given, and salicylate of soda at night. On August 6 she felt much better, the pulse was 80, she slept well, but still had slight pain in the shoulder.

CASE 6.

Chronic Rheumatism.—On June 15, 1896, a married woman aged 61 years came under observation. Her father's mother suffered from rheumatism. Her father died at the age of 49 years from asthma. She had three brothers and three sisters living and well, and had lost three brothers and two sisters, probably some of them from phthisis. The patient had had six children, and her eldest daughter suffered from rheumatism. The patient had

CASE 5.—Subacute rheumatism, taken July 21, 1896, before treatment: showing highest point arm and forearm could be raised, with pain.

CASE 5.—Subacute rheumatism, taken June 21, 1896, before treatment; showing inability to flex fingers on palm.

CASE 5.—Subacute rheumatism taken June 21, 1896, after first bath: showing full extension of arms and forearms and flexion of fingers without pain.

never been confined to bed, but about a year and a half ago she suffered from rheumatism in the knees and other joints, and this had continued ever since. For the last six months it had been getting worse. She had now to be up some hours in the morning before she was able to use her hands even to dress herself. There was no cardiac murmur, but the pulse was rather small. All the fingers and hands were swollen and painful; there was marked wasting of the muscles on the back of the hand. She was given mistura guaiaci. On July 6 the patient was becoming worse; she was unable to raise the left arm and the metacarpo-phalangeal articulations of both hands were more swollen and tender. On the 21st the first bath was given. The left shoulder was very stiff and the arm could only be raised a very little, and she was unable to close the hand. The pulse was very small and feeble. The right arm and hand were placed in the cylinder, and after a short time the pulse much improved in character, and in fifty minutes the left arm could be fully extended, and the hand closed without much difficulty. On the 24th there was still some pain in the left shoulder, but otherwise she was very much better. On the 27th she had had four baths and was able to do a little work. On the 30th she had had six baths. All the pain had gone from the hands and the right shoulder, but she had occasional pain in the left shoulder when in bed and the first thing in the morning. Strychnia and iron were prescribed. On August 6 all the movements of the fingers, hands, and arms were quite free; there was still some pain, or rather what she described as 'numbness in the muscle' of the left arm; she slept well.

CASE 7.

Chronic Rheumatism; duration six months.—A man 59 years of age was seen on July 16, 1896. For six months he had suffered from persistent pain in the right shoulder, which prevented him following his occupation, as he was unable to raise this arm above the horizontal. The pain was worse at night and kept him awake. On April 15 he went into the Greenwich Infirmary, where he remained till June 2. When he came out he was very little better. On July 16 he appeared to be in pain and unable to raise his right arm above the horizontal. A hot-air bath was given. On the 22nd he had had three baths. He reported that he had had much less pain and had been able to do some work. He had slept much better.*

CASE 8.

Treated in the Laennec Hospital, Paris. Under the care of PROFESSOR LANDOUZY.

Subacute Rheumatic Fever; Mitral Affection.—B. T., 23 years old. First (?) attack of rheumatic fever. Mitral obstruction at first unnoticed. Pains generally subacute, causing a relative powerlessness; increased by pressure and by the passive movements. After being treated three times by the superheated dry-air method, the pains were so much lessened that the patient asked to be allowed to leave the hospital.

CASES 9-13.

Treated in the Philadelphia Hospital.

CASE 9.

Thomas Miles; aged 50; birthplace, England; residence, city; occupation, plasterer; attending physician, F. A. Packard; resident physician, Raymond Spear.

Father died of dropsy. Mother died at age of 63; cause of death unknown. Has seventeen brothers and two sisters, all living. No history of rheumatism in family. Has had scarlet fever and measles. Has had two attacks of jaundice. Had great paroxysms of pain at those times. Has used alcohol to excess. Had gonorrhœa twenty years ago. No specific history. Has had frequent attacks of pain in his back. The pain was sharp and shooting in character, very short in duration. These attacks occurred about twice a week. The pain did not shoot down into his groins or testicles or penis. About four years ago was suddenly seized with a sharp pain in his right hip. This pain lasted a month. He kept on working, however. The pain gradually disappeared. He was compelled to stop work for three weeks. He then started to work again. He has been compelled to stop work frequently. The pain was then felt in the anterior portion of his right thigh, also in his back and right knee. The pains were shooting in character and of short duration. Has had frequent attacks in past four years. Last autumn, 1895, was suddenly seized with a sharp pain in his right hip and anterior portion of thigh. Pain short;

* Cases 2-7 published in *Lancet*, August 29, 1896.

followed the course of the obturator and anterior crural nerves. The man walks with a cane and has a decided limp. Legs were swollen on admission, and he passed only a small amount of urine at that time.

On November 19, 1896, was treated by the Tallerman Localized Superheated Dry-Air Process for forty minutes. The temperature was raised to 280°. After the bath the man discarded his cane and walked unaided. He said the pain was all gone except a little in the outer side of his right knee.

On November 20 the improvement in the case continued. The only complaint was a slight uneasiness in the outer side of his right knee.

The apparatus was removed from the hospital, and he could not be given any further treatment.

Case 10.

Timothy Bow; coloured; aged 48; barber by occupation; born in North Carolina; a resident of Philadelphia; father died of cause unknown; mother died of apoplexy; had ordinary diseases of childhood; had pneumonia; had rheumatism nine years ago; had one attack of gonorrhœa; had malaria about five weeks ago. About October 1, 1896, caught cold and had severe pain in back, also his right hip. On admission to the Philadelphia Hospital the pain in his back was intense. He could scarcely walk, and could not even raise himself up in bed the day after admission—in fact, he presented all the symptoms of a violent attack of lumbago. Was admitted on November 3. His condition did not improve. On November 11 he was taken to the College of Physicians and Surgeons, and treated before it by the Tallerman superheated dry-air method, at a meeting of the County Medical Society. The man when lifted upon the bed groaned out with pain. At the conclusion of the bath, which lasted forty minutes, the man got up from the bed unassisted and dressed himself. He said the pain had almost disappeared. He had been carried to the room on a stretcher. He walked down to the ambulance unassisted and into the Philadelphia Hospital, and went to bed unaided. On November 17 the man was discharged cured. He had no return of pain after the hot-air bath. The only complaint he made was that he felt a little weak, which was probably due to his previous confinement in bed and his loss of sleep before the treatment.

RAYMOND SPEAR.

Case 11.

John Hogan; colour, white; aged 48; birthplace, America; residence, Philadelphia; occupation, driver; attending physician, F. A. Packard; resident physician, Raymond Spear.

Father was killed; mother died of old age; had pneumonia on left side twice; had malaria in 1863; had typhoid in 1867; has used alcohol to excess; had a venereal sore in 1866; it was followed by no secondary symptoms; had an attack of rheumatism, involving right shoulder and arm, in 1894; was exposed to weather on September 22, 1896, and caught cold. Immediately following exposure was taken with an acute attack of rheumatism, involving left shoulder. The joint swelled, became red, and was very tender and painful. At that time he could not raise his right arm without great pain. His condition improved up to a certain extent under antirheumatic treatment; then became stationary. The motion in the joint was limited, and the joint was the seat of constant pain. The pain was worse in damp weather. On November 19 was subjected to the Tallerman treatment. After the operation he could raise his arm above his head and the pain had almost disappeared. A second bath was given on November 18, lasting fifty minutes. Temperature was raised to 310° F. The motion in the joint is perfect, and he now has only a slight pain under his left scapula.

Case 12.

Thos. Birmingham; father died of typhoid fever; mother alive; one brother and one sister living and healthy; no history of rheumatism or phthisis; had malaria ten years ago; had scarlet fever and measles and typhoid; had gonorrhœa eight years ago; two years ago had two venereal sores, followed by no secondary symptoms, except a few sores (?) in his mouth. He was told he had syphilis, and took medicine for three weeks. Has had no secondary symptoms since.

In February, 1896, his left knee swelled up and became very painful. His left hip then became involved, then his right elbow. The swelling and pain disappeared from all his joints, except his left knee. This remained somewhat swollen and painful ever since. He could bend his knee only with great pain. The joints felt weak and the motion was limited. On November 17 he was placed under the Tallerman treatment for forty minutes; temperature, 240° to 280° F. After half an hour the pain had greatly lessened and the joint was more movable.

On November 18 was given a second bath, lasting fifty minutes; temperature, 240° to 315° F. The pain almost left the joint, and there was still more motion. On November 19 was given a third bath; temperature, 240° to 280°, with the result that the motion in the joint is almost perfect, and the pain has almost disappeared. On November 20 the improvement was maintained.

CASE 13.

Andrew Cathcart; native; aged 46; single; teamster.

Family History.—Negative as far as present disease is concerned.

Present History.—Disease of childhood; gonorrhœa at 20; used alcohol and tobacco to excess; toes of both feet frosted ten years ago, necessitating amputation of two toes; right ankle deformed from accident, in which a carriage wheel twisted it.

In April, 1893, first noticed that he had pains in the joints of the leg in motion. These began to swell. Then later the tissues became involved; the neck became stiff. This pain persisted, and the fingers became clumsy. The swelling in the joints continued to increase, and infiltration of the tissues about the joint took place, which was greater in the right knee. The joint presented a hard, indurated feeling, rather sensitive to touch, and as the disease progressed there was a progressive loss of motion, due to a fixation of both legs in a semiflexed position. These could not be straightened out, and finally the patient was unable to walk, or even to stand on his feet for more than a minute and a half at a time. There seemed to be a feeling of weakness and 'lack of confidence' in both knees. The right wrist was practically motionless, and the left wrist was nearly so. Both these joints were thickened, infiltrated, and tender to the touch, as were the knees. No cardiac involvement; no pulmonary or hepatic trouble. Persistent partial insomnia, sleeping but one to two hours per night. After Tallerman treatment, two applications of forty minutes each to left leg (November 19 to 20, 1896), patient stated the following morning that he slept well all night, as he has every night since, and that his entire body feels easier, and with a marked diminution of that 'malaise' and feeling of tension which formerly troubled him. The right wrist was slightly reduced in size, but now there was no pain or tenderness. There was an arc of motion of $1\frac{1}{2}$ inches at finger-tips; no lateral motion. The same was true of left wrist, except that the previous range of motion was increased, and now without any pain.

The Knee.—*Left Knee:* The relief here was marked. The joint decreased from $\frac{1}{2}$ inch to 1 inch in all its measurements. The joint was softer. The tissue infiltration was diminished in consistency, and there was lack of tenderness and pain in motion. The leg, which before was fixed at an angle of about 145°, could now, with little effort, be stretched out straight (80°). The patella seemed to be more freely movable. The knee seemed to feel stronger to the patient himself.

Right Knee.—The improvement in this knee was not so marked as the left knee, yet decided and considerable. While the knee could not be straightened out, there was a gain of 15° to 20° of motion. There was here, as in the other knee, a softening of skin and tissues, with a diminution of the consistency and lack of pain and tenderness. There was, however, still a feeling of weakness left in this joint. For the last three days the patient has been able to walk the length of the hospital ward (thirty yards) with the aid of a cane. Previous to this he had to be assisted from his bed to the wheeling-chair by assistants. The improvement in a case of such formidable appearance, and in which we were able to do nothing to check the onward progress of its course, is remarkable.

A. J. McCarthy, Resident Physician,
Philadelphia Hospital.

Case 14.

Treated in South Charitable Infirmary, Cork. Under the care of N. H. Hobart, M.B.

J. K. Had rheumatic fever, September, 1895; laid up for seven weeks. Treated April, 1896. Pain in left ankle and elbow and right heel. Had ten baths. After the tenth he was perfectly well, free from all pain and stiffness, both in foot and elbow, although his elbow had never been inserted in the apparatus.

CASES 15-24.

Treated in the Liverpool Infirmary. Under the care of DR. ALEXANDER.

CASE 15.

Robert D., aged 52, states he has been suffering from rheumatism for three years, and his arms and legs have been stiff and painful ever since the commencement of the attack. Treatment by superheated dry air commenced July 29. This patient had in all thirty-six baths, his highest temperature during bath being 100·4°, while the highest bath temperature was 250°. The bath was changed from arm to leg, and knee to wrist, at the option of the patient. He had to be carried to and from his bath until his nineteenth bath, when he was able to walk back to his own ward. His last bath was on November 19, and he left the hospital on the 21st, quite cured of all pain and uneasiness, but some stiffness of joints remaining.

CASE 16.

Annie R., aged 19, states about two months ago she developed rheumatism in the knees and ankles, spreading soon afterwards to the arms and wrists. She had vapour baths, which had no appreciable effect. Tallerman treatment by bath commenced July 13; had seven consecutive baths, the baths alternating between legs and arms; patient's highest temperature during baths, 99·8°; highest temperature of bath, 225°. Left hospital July 20 quite cured.

CASE 17.

Michael T., aged 33, has been suffering from rheumatism since the beginning of July, 1896, his knees being chiefly affected. Admitted to this hospital August 5, and kept under general treatment till September 17, when superheated dry-air baths were commenced. During early treatment he had to be carried to and from bath. At termination of sixth bath he was able to walk. He had in all sixteen baths, his highest temperature during baths being 100·4°; highest temperature of the bath, 250°. Left hospital September 11, 1896, quite recovered.

CASE 18.

Isaac S. was always healthy until June, 1896, when he had a severe attack of rheumatism. Was in hospital abroad for three weeks, and, feeling somewhat better, came to Liverpool. Admitted to this hospital August 7, suffering from a fresh attack on top of the old. Treated with sod. salicyl., vapour baths, joints painted with iodine liniment, but patient did not seem to get any better. Began Tallerman baths August 20, 1896. Patient had fourteen baths, was carried to first, second and third baths, but afterwards able to walk; his highest temperature in bath 100°; highest temperature of bath 245°. His last bath was September 5, when he was free from all pain, stiffness only remaining. He was found dead in bed October 24, without any previous complaint.

Case 19.

John C.; aged 35; admitted to this hospital October 19, 1896, suffering from rheumatism in right hand. His first bath was on November 5, and the last November 19, having had four baths. Highest temperature in bath 99·8°; highest temperature of bath 250°. Left hospital November 21, 1896, quite recovered.

Case 20.

Robert B., aged 45; admitted to hospital October 27, 1896, suffering from rheumatism, having had a previous (and only) attack in the autumn of 1887. Rheumatism confined to hands and knees. Baths commenced November 17, and ended December 14, having had ten baths in all. Patient's highest temperature in bath 100·8°; highest temperature of bath 260°. Left hospital January 6, 1897. with only an occasional twinge of pain in hand.

Case 21.

Thomas M.; aged 37; admitted to hospital, suffering from rheumatism of both knees for twelve years. Had soda lotion, etc., for three weeks, but only a slight improvement. Had only four baths when he expressed himself cured. Patient's highest temperature in bath 100°; highest temperature of bath 220°. Left hospital December 8, 1896.

Case 22.

John E., aged 42, states that he is suffering from rheumatism of right leg, and has been so for two and a half years. Admitted to hospital April 30, 1896. Bath treatment commenced July 28, finished November 14, having had twenty-two baths. His highest temperature in bath was 101·2°; highest temperature of bath 235°. Left hospital soon after last bath, feeling much easier and able to walk a great deal better since commencing bath.

Case 23.

Edwin B., aged 48, states he had a sharp attack of rheumatism when 26 years of age ; six years elapsed before a second attack, and he has never been free from rheumatism since. Was treated at Bath Mineral Water Hospital, and after six months was discharged apparently well. Has been in this hospital three or four times, always with rheumatism. Re-admitted July 30, and Tallerman bath treatment commenced August 17, 1896. He had eight baths, and expressed himself greatly improved by treatment. Patient's highest temperature in bath 100° ; highest temperature of bath 256°.

Case 24.

Joseph J. C., aged 54, states he has suffered from rheumatism since boyhood. Admitted here June 26, suffering from rheumatism, principally of hands. Baths commenced August 29, and finished September 5. Had five baths ; fingers better—able to bend them freely. Patient's highest temperature in bath 99·4° ; highest temperature of bath 235°.

JOSEPH MAGUIRE, Resident Medical Officer.

Case 25.

Treated at the Royal Portsmouth Hospital. Under the care of MR. D. WARD COUSINS, F.R.C.S.

I. C.; aged 24. Carpenter. Invalided from Royal Engineers for chronic rheumatism. Has been unable to work for some months on account of pain in right wrist. Considerable thickening round wrist-joint. Range of motion limited and grasp very feeble. Attempts to move joint caused considerable pain.

First Operation, October 18. *Forty minutes;* 240° *F.*

Hand and arm treated by superheated dry air. After twenty minutes, patient stated that he was quite free from pain.

Second Operation, October 20.

Patient states that since the last operation he has suffered much less pain. The grasp is stronger. Range of motion increased, and the pain accompanying it is less. Placed hand in cylinder for forty minutes at 240° F. Thickening round joint decreasing.

Third Operation, October 23.

The thickening round the joint has much decreased. Can now grasp firmly with the hand, and is almost free from pain. Hand placed in cylinder for thirty minutes at 260° F.

Fourth Operation, October 25. *Forty minutes;* 260° *F.*

Thickening almost gone. Pain very slight. Patient was, unfortunately, not able to attend any more. The improvement, however, was most marked, the thickening being all absorbed, the range of movement and strength of grasp being considerably increased, and when seen on November 2 he could flex and extend the wrist-joint almost to the full extent without any pain, whereas before the treatment the movement was limited in extent and accompanied with severe pain.

T. H. BISHOP, M.B., C.M., House-Surgeon.
H. W. MORLEY, M.R.C.S., L.R.C.P., Assist. H.-S.

CASE 26.

Treated at the Livingstone Cottage Hospital, Dartford. Under the care of T. F. CLARKE, M.D.

Muscular Rheumatism.—F. B., aged 20 years, has suffered off and on from muscular pains for the last five or six years. These pains have not been confined to any one part of the body, but have been more or less general.

The present attack commenced about four weeks ago, the pains being confined to and between the shoulders.

The occasional exacerbations of pain are very severe, and begin to affect the general health.

On March 28, 1896, the right arm was placed in the hot-air bath for forty minutes, the temperature of the bath reaching 250°. After the application she felt quite free from pain for the next twenty-four hours, and even then the pains were very slight to what they had been before.

Further applications of the hot-air bath were made on March 31, and April 4, 8, 11, and 13. The pains did not altogether disappear: they would cease for a day or two and then return, but to a very limited extent.

If still further applications could have been made I feel convinced that the rheumatism would have soon disappeared altogether, but the patient was obliged to leave Dartford.

July 24, 1896.—This was the last occasion on which I met Miss B., when she told me that a day or two after the last application of the bath in April the pains entirely disappeared, and had never returned since; and that she had never experienced immunity from pain for so long a period since she was first attacked with rheumatism some five or six years ago.

CASE 27.

Treated at the same Hospital as Preceding Case.

Chronic Rheumatism.—E. G., aged 64 years, has a very rheumatic history, both parents having been great sufferers. Her first attack occurred twelve years ago, when all the joints on the right side were affected. Since then she had been laid up every year, but in 1894 she had two severe attacks.

At the present time most of the joints are more or less chronically enlarged. She had been unable to do any work for the last few months on account of the increased swelling and stiffness of the right shoulder-joint, not being able to raise the arm to a right angle; and also on account of the right elbow-joint being stiff and causing much pain on movement.

On March 18 the patient's right arm was placed in the hot-air cylinder for forty minutes, the temperature of the bath reaching 280°. On the limb being removed from the bath the patient was able to extend and flex the elbow to the fullest possible extent without the least pain, and also to raise the arm well above and to the back of the head without pain.

The application was repeated on March 22 and again on March 25, and after this the patient was enabled to do field work, a thing she had not done for a considerable time.

It was thought advisable to persevere with the hot-air treatment occasionally, so she had further baths on March 31 and April 11; the free and painless movements of the joints, however, continued in every way satisfactory.

September 5, 1896.—About the middle of May this patient had a return of the pain and swelling in the right shoulder and elbow joints, brought about from doing hard field work in wet weather. Under simple ordinary treatment the symptoms soon abated, and she was again able to resume work. I need hardly remark that if the hot-air bath had been at hand I should have immediately made use of it; but this case certainly points to one important fact—viz., that the good results following the application of the hot-air bath are more or less permanent.

CASE 28.

Treated at St. Bartholomew's Hospital. Under the care of MR. WILLETT, F.R.C.S.

The next case in which the bath was used was that of a woman about 45 years of age, who had been one of my in-patients twelve months previously for acute rheumatic inflammation of the wrist-joint. Her attack had been a very severe and prolonged one, and the whole hand and forearm had been throughout kept at rest. Eventually the patient recovered. But she came recently complaining of stiffness in her hand. Flexion and extension, pronation and supination of the wrist-joint were very good. Active mischief in the wrist-joint had entirely passed off, but secondary stiffness of the fingers had resulted. She had the bath treatment for two months very regularly, but I must own that the result of its use in this case greatly disappointed me. I regarded it as the best test case of the series, for her condition was that of a rheumatic joint which had entirely recovered, leaving only stiffness of fingers and hand, resulting from the long-continued immobility. It seemed to me to be exactly the kind of case to be cured. The original affection may, however, I think, have been more diffused and extensive than had been diagnosed, and so perhaps, in addition to synovitis of the wrist-joint, there had been teno-synovitis of the sheaths in front and behind the wrist. But from whatever cause, I cannot doubt that fibrous adhesions within the sheaths had resulted, or that this accounted for the little benefit that was observed after two months of treatment. I have seen that patient within the last fortnight, and I am bound to confess that she seems not in the least degree benefited by the prolonged bath treatment.*

This is an interesting and instructive case, on account of the failure of the treatment and the explanation suggested by Mr. Willett.

* *Clinical Journal.*

Case 29.

Treated by A. ROBERTS, M.D.

Chronic Rheumatism.—Mrs. F., aged 35, had swollen wrist and fingers for two years, since last confinement, but had not been treated medically for the rheumatic joint.

Patient was unable to wring clothes with her hands, and they were very painful to the touch; there was little or no power of grasp.

The right hand was placed in the Tallerman apparatus at a temperature of about 150° and raised to 230°. The left hand that was outside the cylinder was carefully examined by the medical gentlemen present, and found to be stiff, swollen and painful, similar to the right.

The operation extended over forty minutes, with the result that all the pain had gone from both hands and the swelling of the wrists very much relieved.

Mr. Tallerman's engagements compelling him to return to town, he was induced to leave the apparatus behind, in order that the patient might be further treated. Next day the improvement was found to be maintained, and she was able to knead her bread and cake and wring clothes without pain, which she had been unable to do for some time. There was also considerable power of grasp. A week later she was treated for the second time, and for the third and last time a few days later. She is now practically cured. The cure was effected by the bath treatment solely, as, this being a test case, not any medicine or other treatment was used.

Case 30.

Treated at 50, Welbeck Street. Under the care of DR. CROSS, B.A., M.D.

Mrs. C.; aged 54.

History.—In August, 1893, the disease first developed, and patient was ill for three months; the knees, wrists and elbows were swollen and painful. This was followed a few months later by a slight attack, and in February, 1896, there was a severe attack, during which all the joints became affected. Patient was at this time an inmate of Middlesex Hospital; she was subsequently sent to a Convalescent Home at Brighton, where she derived great benefit from bathing in hot sea-water. On April 20 and 24 patient again consulted me, her hands being still swollen and painful. I saw her again at the beginning of August, and advised her to try the Tallerman treatment, as her hands continued painful and comparatively useless, the shoulders were so stiff and painful that she was unable to raise her arms above her head, while the knees were painful and swollen, preventing her walking any distance. Patient attended at the Tallerman Institute from August 10 to 18.

On examination, August 10, 1896, it was found that the right shoulder was extremely painful and movements limited; hands swollen, stiff and very painful, unable to flex the fingers of either hand to the palm, and there was a lump on the top of right wrist; knees swollen and painful when walking and in going up and down stairs. Considerable pain in lower part of back. Patient has great difficulty in lifting or carrying anything, and has bad nights, the pain preventing her from sleeping.

First operation, August 10, 1896.—Right arm treated; the movements, after treatment, of the arm and shoulder were normal and free from pain, and patient could close both hands (she had not been able to do this for many months); the lump on the wrist was very much smaller, and the swelling over the knees had decreased.

After the fourth operation, August 18, 1896, patient reported that she was now quite free from pain, that the hands and arms were much stronger, and that she could lift and carry quite well; she can go quickly up and down stairs without pain; the lump on the wrist is hardly visible; she sleeps well and feels much better in herself. It was therefore thought unnecessary to treat her further.

Patient was treated at an average temperature of 240° F. for forty-five minutes, the body temperature being raised 1°, and the pulse beat increased from 12 to 16.

The skin acted profusely during each operation.

Mrs. C. called on me on August 28, and stated that she still kept free from pain. I found that she could flex both hands perfectly and raise her arms well above her head without pain, and the weakness of the legs had disappeared.

Case 31.

Treated at the Tallerman Free Institute.

Chronic Rheumatism.—Mr. W.; aged 60. Pensioner. The disease commenced six years ago, attacking arms, elbows and shoulders, and finally settled in left ankle and foot. Patient had undergone treatment at Guy's, St. Thomas's and Charing Cross Hospitals.

On examination, the left ankle and foot were considerably swollen and limited in movement, and there was extreme pain and difficulty in walking.

First operation, April 14, 1896.—There was great pain in the ankle-joint before it was placed in the cylinder, but after treatment patient could walk with very little pain.

After third bath it was noted that the ankle was reduced to normal, and patient reported that all his joints were quite free from pain. The treatment was continued.

Eighth operation.—The improvement having been confirmed, and complete mobility of the joint restored, and all pain having subsided, patient was discharged.

Average time of each operation, fifty minutes.
Average temperature attained, 230°.

Body Temperature.		Pulse Beat.	
Before Operation.	After Operation.	Before Operation.	After Operation.
98·0	99·2	68	80
98·4	99·2	72	88
98·2	99·2	80	80
98·0	98·6	72	76
98·4	99·2	76	96
98·4	99·4	76	84
98·4	99·4	68	80
98·8	99·2	72	88

Case 32.

Treated at the Tallerman Free Institute.

Chronic Rheumatism.—M. P.; aged 58. Housewife. Has suffered on and off for five years. There was pain throughout both legs, the left leg being the worse. The knees and ankles of both were very swollen and extremely painful.

The hands, especially the fingers, were somewhat affected, being painful at times; there was no swelling. At one time patient had been unable to walk, but could walk now with a little assistance.

First operation, April 28.—After treatment, and on measurement, the knee was found to have been reduced half an inch, and patient could move it with much less pain.

After third bath, patient reported that she had been able to kneel, which she had not been able to do for five years, viz., since the commencement of the disease. She also said that she was entirely free from pain.

Notwithstanding the improvement, it was thought advisable to continue the treatment, as there was some slight contraction of the left leg.

After tenth operation it was found on measurement that the knee had been reduced 1¼ inches, and that with the slightest possible pressure the leg could be fully extended.

Patient reported now that she could walk without assistance, felt stronger, and that her general health had been greatly benefited.

Average time of each operation, forty-five minutes.
Average temperature attained, 250°.

Body Temperature.		Pulse Beat.	
Before Operation.	After Operation.	Before Operation.	After Operation.
98·4	99·4	88	100
98·4	99·6	92	102
98·8	99·0	92	106
99·0	100·0	88	96
98·6	100·0	92	104
98·4	99·8	88	96
98·4	99·8	92	104
98·0	99·6	88	96
98·2	99·2	92	106
98·6	99·2	84	96

CASE 33.

Acute Rheumatism.—J. O.; aged 20. Occupation, clerk.

Family History.—No phthisis. Mother suffered from rheumatism.

Previous History.—No serious illness; always enjoyed good health.

Patient's account of present illness: On December 1, 1896, whilst writing noticed his right hand becoming painful, hot, and swollen; left off writing and placed it under cold water tap for a few moments. Next morning finding hand still swollen and painful, saw Dr. Carter, who diagnosed 'Rheumatism,' and treated him with salicylates and salines. At the end of fourteen days, finding his hand no better, he went to St. Mary's Hospital; there he was told that he was suffering from tubercular disease of the wrist-joint; his hand was put in a splint, belladonna applied, and iron given internally. Patient was recommended to go into the country for a few weeks. At the end of one month splint was removed, the wrist and fingers were stiff, but the pain was gone; at the end of two months he returned to work, he was able to hold his pen and, with difficulty, write. After a few weeks the pain and swelling returned.

June 20, 1897.—Patient consulted me in respect to his hand. On examination I found it hot, perspiring freely, extremely painful and swollen. The wrist-joint could neither be flexed nor extended, the metacarpal and phalangeal joints allowed but very slight movement. Patient stated that after writing a few minutes his fingers became so painful and stiff that it was impossible to continue.

Having recently seen very great relief and good done by the use of Mr. Tallerman's superheated dry-air baths in a case of acute gout, I determined to ask Mr. Tallerman to undertake this case, which he very kindly did.

The following table shows the progress of the case whilst undergoing Mr. Tallerman's treatment :

Date.	Temperature of Bath.	Duration.	Remarks.
July 3	290°	45 minutes	On withdrawal, no pain in hand, even upon full manipulation.
July 5	280°	1 hour	Some increase of movement in wrist. Fingers less painful when writing.
July 6	280°	1 hour	Increase of movement.
July 8	280°	1 hour	Patient says he has had no pain since last bath. No swelling.
July 9	280°	1 hour	No return of pain. Fingers much more flexible.
July 12	260°	1 hour	Patient can flex fingers to palm himself; increase in wrist movement.
July 13	250°	1 hour	Patient has been writing a good deal, and wrist is stiffer.
July 16	260°	1 hour	Patient has given up situation.
July 19	260°	1 hour	Fingers and wrist decidedly more flexible.
July 20	260°	1 hour	No return of swelling or pain; free movement in fingers.
July 22	260°	1 hour	Ditto ditto.
July 23	260°	1 hour	Ditto ditto.

On July 25 I saw the patient, and found that there was normal movement in metacarpal and phalangeal joints, but that there were evidently some adhesions in wrist-joint; all pain and swelling had disappeared.

Remarks.—It must be admitted that the effect of these baths upon this case has been extremely satisfactory, and I have no doubt but what, had they been applied earlier, the still remaining slight stiffness in the wrist would have been prevented, and we should have had as free movement in this joint as we now have in the fingers. The almost immediate removal of all pain is very marked, and renders this form of treatment extremely valuable, as it is also from a diagnostic point of view. In this case there was some difference of opinion as to whether it was tubercular or rheumatic; the result, after treatment, leaves no doubt but what it was a case of simple rheumatism.

E. CHINSON GREENWOOD, L.R.C.P., M.R.C.S.

CASE 34.

Treated at the Tallerman Free Institute, Blackfriars.

Subacute Rheumatism — Cure in One Operation.— D. L.; aged 41. Builder's labourer. Patient has been a sufferer from rheumatism for fourteen years, and has been treated at St. Bartholomew's and St. Thomas's Hospital with very little benefit. He has lost nearly four months' work this year. The last attack was in the right wrist, when he was unable to work for seven weeks. During a thunderstorm on Monday, July 19, it came on again in the left ankle. The pain was so great that he could not bear to put his foot to the ground. In this condition he came to the Institute on Friday, July 23, on a coster's barrow, as shown in the photograph. After one operation he was able to put his shoe on and walk without pain. He returned to his work on the following Monday.

Duration of bath, 60 minutes.

	Pulse.	*Body Temperature.*
Before	60	99°
After	96	100·4°

October 14.—He has remained at work ever since. His wife states that he is quite free from pain, and was never better in his life.

Case 34.

Case 34.—After treatment.

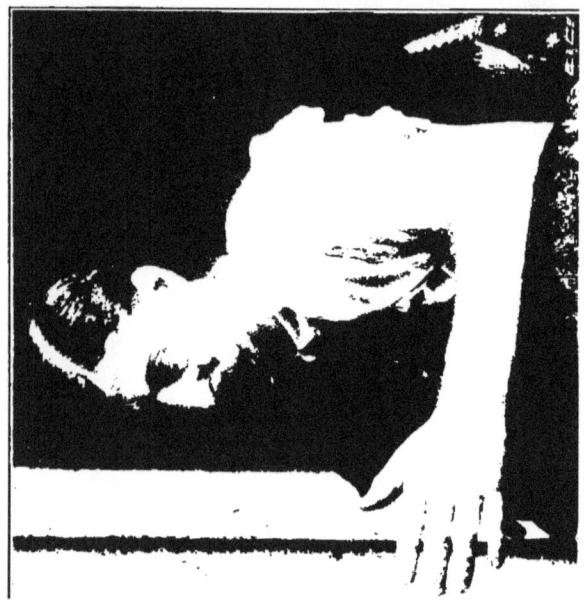

CURE OF ACUTE RHEUMATISM TREATED BY DR. CHRÉTIEN OF PARIS.

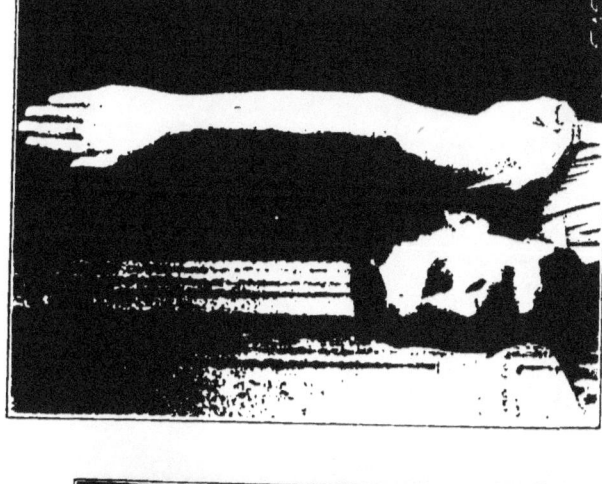

After treatment. June 16.

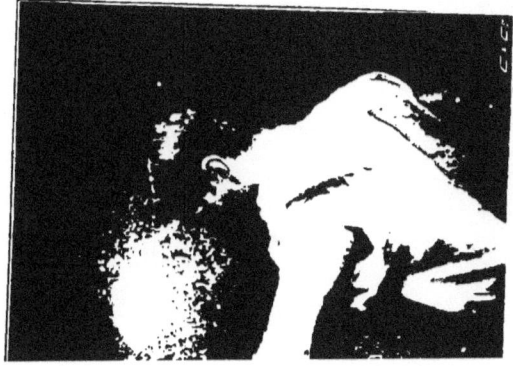

treatment. June 1.

CHAPTER IV.

GONORRHŒAL RHEUMATISM.

THE following eight cases show that the hot-air treatment is remarkably effective in this troublesome disease. (See also Reports by Dr. Dejérine and Professor Landouzy, p. 145.) In mild cases it promotes an extremely rapid cure; in severe ones it cuts short the mischief and prevents permanent disablement. Mr. Willett's case is particularly striking. The man was getting into a very bad condition, and no improvement took place until he was subjected to the Tallerman treatment, under the influence of which he rapidly recovered. In view of recent disclosures concerning the health of the Indian army and the large number of men disabled by gonorrhœal rheumatism, a method of treatment both rapid and effectual deserves the serious attention of the military authorities. Why is it not tried at Netley Hospital?

CASE 1.

Treated at St. Bartholomew's Hospital. Under the care of
MR. WILLETT, F.R.C.S.

This case was that of a man named James L., 26 years of age, a plumber, suffering from urethral synovitis. One month he has had discharge, for two weeks and a half pain in the left knee, afterwards in the right ankle and both feet. On admission his left knee, right ankle and both feet were swollen, tender and hot, but there was no marked redness. There was no cardiac murmur and no sweating. The temperature was raised 2° to 3°. Past history: Rheumatic fever twenty years ago. During the first two weeks after admission he had much pain in the knees, ankles and feet continuously, but varying in in-

CHAPTER IV.

GONORRHŒAL RHEUMATISM.

THE following eight cases show that the hot-air treatment is remarkably effective in this troublesome disease. (See also Reports by Dr. Dejérine and Professor Landouzy, p. 145.) In mild cases it promotes an extremely rapid cure; in severe ones it cuts short the mischief and prevents permanent disablement. Mr. Willett's case is particularly striking. The man was getting into a very bad condition, and no improvement took place until he was subjected to the Tallerman treatment, under the influence of which he rapidly recovered. In view of recent disclosures concerning the health of the Indian army and the large number of men disabled by gonorrhœal rheumatism, a method of treatment both rapid and effectual deserves the serious attention of the military authorities. Why is it not tried at Netley Hospital?

CASE 1.

Treated at St. Bartholomew's Hospital. Under the care of
MR. WILLETT, F.R.C.S.

This case was that of a man named James L., 26 years of age, a plumber, suffering from urethral synovitis. One month he has had discharge, for two weeks and a half pain in the left knee, afterwards in the right ankle and both feet. On admission his left knee, right ankle and both feet were swollen, tender and hot, but there was no marked redness. There was no cardiac murmur and no sweating. The temperature was raised 2° to 3°. Past history: Rheumatic fever twenty years ago. During the first two weeks after admission he had much pain in the knees, ankles and feet continuously, but varying in intensity. His joints were all stiff. His temperature was raised usually, frequently to 101°. He was relieved by blisters. The movements of the joints improved.

About February 20 he was placed in the hot-air baths. From that time his improvement commenced. He left the hospital at the end of March. The movements of his joints were greatly improved, and all active inflammatory condition passed away. His knees and ankles and right shoulder were slightly painful, but the movements were nearly natural. This was an extremely severe case of that form of rheumatism, and one which, I think, might have left him crippled by adhesions, although he would have got well in time; yet no improvement took place until one ankle and knee were placed in the cylinder, and from that time convalescence commenced. I feel convinced the bath treatment greatly accelerated recovery, and probably prevented some serious organic changes in the joints. That patient was five weeks under treatment.*

The accompanying photographs show the completeness of the cure.

CASE 2.

Treated at the Royal Victoria Hospital, Montreal. Under the care of PROFESSOR JAMES STEWART.

GONORRHŒAL ARTHRITIS OF THE RIGHT KNEE OF FOUR WEEKS' DURATION—INABILITY TO WALK OR EVEN STAND—AFTER THE THIRD BATH HE WAS ABLE TO WALK ABOUT THE WARD WITHOUT ASSISTANCE—AFTER THE TWENTIETH BATH HE WAS DISCHARGED FREE FROM PAIN, AND WITH GOOD MOVEMENT IN THE JOINT.

J. L., aged 22, was admitted to the hospital on December 18, 1896, complaining of pain and swelling in the right knee, with inability to walk.

The history of the case is briefly as follows: In October,

---

* The following letter was afterwards received from this patient:

'111, Rosoman Street, Clerkenwell, E.C.,
'July 1, 1894.

'DEAR SIR,—I take the pleasure of writing to let you know that I am the patient mentioned in the *Clinical Journal* by Mr. Willett, in his lecture—J. L.; aged 26; plumber; and since leaving the hospital the pains in my shoulders and knees and ankles have all left me, and I thought right to call at the hospital and let Mr. Willett see me, and I feel now as strong and as well as ever I was; and as Mr. Willett stated, I might have been crippled by adhesions, or probably suffered from serious organic changes in the joints, I offer my sincere thanks for the treatment of the bath, and if anybody wishes, they can see me at any time.—I remain, yours truly, 'J. LATHAM.'

CASE 1.—After treatment: showing free and painless movement.

To face p. 72—1.

Case 1.—After treatment: showing free and painless movement.

To face p. 72—II.

CASE 1.—After treatment: showing free and painless movement.

CASE 1.—After treatment: showing free and painless movement.

To face p. 72—IV.

1896, he contracted gonorrhœa, and up to the middle of November had a profuse urethral discharge. At this time the knee suddenly became much swollen and intensely painful; the conditions lasting about three weeks, when it became less painful, but was still swollen and very stiff. There was no constitutional disturbances other than rapid loss of flesh. Of his personal history nothing was ascertained save that he had used alcohol and tobacco to excess.

On admission, the patient was found to be unable to stand without support, and could not walk at all. The knee could be flexed and extended but a short distance, while there was uniform enlargement of the tissues about the joint, but no increase in the synovial fluid. No other joint showed abnormally. The urethral discharge contained gonococci.

The treatment, as in the previous case, was the Tallerman superheated dry air. After the first application a decided improvement was noticed, while after the third the patient was able to get about the ward without assistance. In all he had twenty applications, and on discharge, January 27, 1897, there was free movement of the joint and practically no pain in walking. Although there still remained some enlargement and deformity about the joint, his general health had much improved, as evidenced by a gain in weight of 14 lb. during the period of treatment.

CASES 3-5.

Treated at the Laennec Hospital, Paris. Under the care of PROFESSOR LANDOUZY.

CASE 3.

Gonorrhœal Rheumatism. — Bertha B.; 16 years old. Gonorrhœa, followed by diseases of the large joints. Watery blisters on the soles, exceedingly painful to touch and rendered walking difficult. The right knee only was affected; there was pain, flexion almost to a right angle, and functional impotence. After eleven days and without good result, the Tallerman treatment, superheated dry air, was tried. At the end of three operations the pain had completely disappeared; the knee will be fully extended again. The appearance of pain in some of the joints necessitated the continuance of the treatment. At the end of ten baths the patient was completely cured and left the hospital.

CASE 4.

Gonorrhœal Arthritis of Several Joints; Talalgia.—M.; 27 years old. Severe gonorrhœa, on the third day of which severe pains appeared in the left heel, followed by general pains in the knee and right elbow. Severe talalgia. Unable to walk. After the second bath the pain had completely disappeared, both in knee and heel, and the patient walked about all the afternoon. After the third bath the patient left the hospital of his own accord.

CASE 5.

Infectious Arthritis (Gonorrhœal?) of the Right Wrist.— Louis J.; 36 years old. Several attacks of gonorrhœa, but at long intervals. The actual illness, which was not coincident with any urethral incident, began in a painful attack of rheumatic fever, which affected exclusively the right hand and wrist. This part was attacked by a moistness, an intense swelling, hard, inflamed, and slightly painful under pressure. The movements of the fingers were scarcely noticeable. The flexion and extension of the hand and wrist were entirely impossible. Slight atrophy of the forearm. This joint disease of a somewhat intermediate nature had resisted all therapeutic methods for about three months. We then tried the superheated dry-air treatment.

After the eighth bath the swelling was considerably diminished. The fingers and wrist regained their flexibility; so much so that the patient was able to write a four-page letter, and write it well.

CASE 6.

Treated at Charing Cross Hospital. Under the care of MR. J. H. MORGAN, F.R.C.S., Sen. Surgeon to the Hospital.

H. J. M.; aged 23. Postman. Second attack of gonorrhœal rheumatism in knees, ankles, and wrists. After various treatments had been tried, the superheated dry air was used for the right knee and left wrist. This treatment was commenced on February 6, and was continued daily, the knee being treated until the 14th, when the wrist also was treated. Patient left hospital on March 16, and his cure was apparently much accelerated by the bath.

CHARLES GIBB, Surgical Registrar.

CASE 7.

Treated at Charing Cross Hospital. Under the care of
MR. WATERHOUSE, F.R.C.S.

A. B.; aged 24. Gonorrhœal wrist. Terminating in rapid cure.

CHARLES GIBB.

CASE 8.

Shown at the Brighton and Sussex Medico-Chirurgical Society by DR. HEDLEY.

Subacute Articular Rheumatism ('Urethral Synovitis'); a second attack.—Patient can walk tolerably well, but is very perceptibly stiff. Articular symptoms are of six weeks' duration. Knees chiefly affected, right somewhat swollen, painful on movement or pressure. Left slightly so, ankles similarly affected in a less degree, pain under tendo Achilles. Right knee placed in a Tallerman cylinder at a commencing temperature of 160° F., gradually increased to 240° F. Duration of operation forty-five minutes. Repeated in two days. On termination of third such procedure, the patient volunteered the statement that it was 'the better leg of the two.' Though this relative improvement did not last, an actual improvement did. After a fifth operation, fifteen days from the first, it was not considered necessary to continue the treatment, inasmuch as the symptoms had nearly disappeared, the improvement being apparent not only in the limb directly acted upon, but in all the affected joints. A general diaphoresis had in each instance accompanied the local application. It is to be remarked that the symptoms were never severe, and that before the hot-air applications they were already improving under other treatment; but it seems tolerably evident that the progress of cure was at least hastened by the measures in question.

CHAPTER V.

STIFF, PAINFUL AND ANCHYLOSED JOINTS — TUBERCULOUS AND TRAUMATIC INFLAMMATION — SYNOVITIS WITH EFFUSION—STIFFNESS AFTER FRACTURES AND OTHER INJURIES—PERIOSTITIS.

TWENTY cases of inflammation, other than rheumatic or gouty, and of injury, resulting in pain, stiffness, and anchylosis, are here given. It will be seen that the pain always yields to the treatment, and stiffness in a great measure. Where fibrous anchylosis exists, greater freedom of movement is often gained without forcible measures; and when those are adopted the subsequent progress of the case is much more rapid than it would otherwise have been. In simple cases of acute inflammation absorption is greatly promoted.

CASES 1-6.

Treated at the Hospital for the Ruptured and Crippled, New York. Under the care of V. P. GIBNEY, M.D.

CASE 1.

Firm Fibrous Anchylosis after Tuberculous Ostitis of the Knee.—A boy 10 years of age was admitted to the hospital in August, 1890, with a knee extensively suppurating, and one that required a partial arthrectomy as an immediate measure of relief. He was discharged in July, 1893, with two sinuses discharging and a range of 15° of motion. His angle of greatest extension was 175°, lacking 5° of being perfectly straight.

He was readmitted in June, 1896, with his knee locked at an angle of 140°—that is, nearly midway between a right angle and a straight line. The sinuses had closed. Under traction his deformity was overcome by repeated stretchings, frequently with an anæsthetic. For two or three months he has worn no brace, except a perineal crutch.

On November 27, his knee could be extended to 180°, and could be flexed to 176°. He was placed in the Tallerman apparatus; pulse, 92; temperature, 98·2° F. He remained in the bath forty minutes. The highest temperature reached was 225° F. At this time his pulse was 116; temperature, 99·6° F. On removing his limb from the cylinder, it could be easily flexed to 168°, which was a gain of 4°.

November 29.—Second bath. Temperature, 98·2° F. Time in apparatus, fifty minutes. Highest temperature, 250° F. A gain in motion of 4°.

November 30.—Highest temperature, 260° F. Under treatment fifty minutes. Gain of 3° in motion.

December 1.—Highest temperature, 262° F. Forty minutes in bath. Gain of 3°.

December 2.—Body temperature, 98·8° F. before treatment; immediately after treatment, 100·4° F. The bath was 260° F. Knee flexed to 158° before, 155° after. Duration, fifty minutes.

December 3.—There was a gain of nearly 2° F. in body temperature. The bath itself was 260° F., forty minutes in duration. A gain of 7°.

His seventh bath occurred to-day. Gain of 1° in body temperature. Highest temperature in apparatus, 280° F.

He has worn no apparatus since the treatment was begun, has used the limb, and, while his range of motion has not increased over 20°, the motion is much more easy. He has been free from pain, and there has been no reaction whatever from the use of the limb. His would seem to be an excellent case for very slight motion under an anæsthetic, followed immediately by the use of the bath.

CASE 2.

Fibrous Anchylosis in a Tuberculous Knee.—No suppuration.—A boy, 7 years of age, admitted June 12, 1896. His disease dates from August, 1895. He has worn a protection apparatus since June. There was, at the time he came under treatment, no sign of abscess. On July 29 it was noted that his leg was perfectly straight, with a small range of motion—say 15°. No signs of active disease. His case was selected as one for experiment, although it was very doubtful whether the active disease was fully arrested. I was desirous of learning the effect upon a knee of this kind. At the date of his first treatment, November 27, his knee could be moved over a small arc, 5°. The bath lasted fifty minutes; temperature, 250° F. Without any force there was a gain of about 3° in flexion. His bodily temperature was raised 1·5° F. at the fourth bath, 1° F. at the fifth, and a little less than a degree at the sixth. The highest temperature of any one bath was on this date, and it reached 280° F. There has been a gain of about 12°. No force has been employed, no apparatus used, and, while the knee presents a little fulness, he has had no pain.

CASE 3.

Anchylosis of Deformed Knee, Result of Tuberculous Disease.—A girl 14 years of age was admitted to the hospital in October of the present year, with a history of tuberculous disease extending over a period of six years. She has had two years of bed treatment. At the time of her admission it was impossible to extend the leg beyond 130°. There was no motion allowed at the knee, and any efforts to move the joint elicited great pain. We made an attempt to correct by means of splints, and succeeded in gaining a few degrees, but the treatment was very painful, on one or two occasions requiring an anæsthetic. After two or three baths were given in the cylinder, we found that no gain was made in the correction of the deformity, and that she was just as sensitive if any attempt at motion was made.

On December 3, under nitrous oxide, I moved the knee over an arc, say, of 15° or 20°, bringing it into a better position, and at the same time breaking up a number of adhesions. Coming out of the gas, she complained bitterly of pain, and the knee was at once put into Tallerman's apparatus; the highest point reached was 280° F. It was fully ten minutes before she got relief. The temperature of the bath on entering was about 180° F. It required 100° F. more to give her the necessary relief. A simple posterior splint was applied on removing the limb. She did not require an anodyne during the night, and her night report was very good. It may be safely said that the excruciating pain which follows an operation of this kind was lessened by the use of the Tallerman apparatus. It would have been better to have put the limb in at the temperature of 280° F., and then raised the heat to 300° F.

These three cases of tuberculous bone disease in children may be taken as fair samples of the good effects which one may expect from the superheated dry-air treatment.

The case which I am about to report now, I think, illustrates its advantage in convalescent cases of hip-disease.

CASE 4.

Subacute Arthritis of Knee, Traumatic, with a Small Range of Motion and Pain on Use.—A medical gentleman, 27 years of age, in January, 1893, while skating, fell with knee sharply flexed. From that time up to this present report he has been under the care of Dr. Lewis A. Sayre, who has made use of various appliances; and in addition to Dr. Sayre's care, the patient has subjected himself to various methods of treatment at the hands of different physicians without obtaining a cure. From time to time he has been nearly well, and would then get a fall or an injury of some kind, which would bring on relapse. At the time he came under the Tallerman treatment (December 1) he had a range of motion from 10° to 15°; but beyond this limit the knee pained him a good deal. There were no infiltration, very little tenderness on handling, and no bony enlargement. He was on crutches, wearing a splint, anterior and posterior. He has had four treatments by the hot-air bath, and at the expiration of every treatment he feels much better, is able to use his limb with more ease, and the range of motion is just a little increased. He expresses himself as very much encouraged, because he can use the limb more freely about his room, can bear his weight on it with less pain and distress than heretofore, and it would seem that an excellent result can be obtained.

CASE 5.

Convalescence in Tuberculous Ostitis of the Hip in an Adult: Joint Free from Deformity, but quite Stiff and Painful on Use.—A gentleman, 22 years of age, was placed under my care by Dr. Robert F. Weir on November 25, 1887. He was hearty and robust-looking, was an athlete in college, and four years prior to this date he had sprained his hip while playing baseball. There was a little stiffness at first, but this soon passed off. It was a year later before any signs of discomfort appeared. He then found it difficult to participate in his exercises, and rest was enjoined. At the time I saw him he was just recovering from an attempt at correction of deformity and removal of adhesions. A diagnosis of ostitis of the hip was readily made, and he was placed under protection treatment. This has continued to the present time. Abscess has formed, been opened and finally closed; a second one appeared and then disappeared. In the last two years I have endeavoured to induce him to go without his apparatus, but whenever he made the attempt pain would be excited, he would rest poorly at night, and he had been quite loath to do away with the traction. He has had the benefit of the best climate, and no pains have been spared to keep his health in excellent condition.

December 3.—He expresses himself after the third bath, now, as very much encouraged. Slept better last night than he has for a long time. The limb can be held much more freely than I have ever seen it. He has not been using his brace since November 30. He is bearing his weight on the limb and walking about the floor without any discomfort.

December 4.—At the meeting this evening he states that he has less dread and apprehension of using his limb. He finds more security when he steps upon it; can lie in bed and swing the left limb around, with leg extended, in a pretty good range, without fatigue or pain. The range of motion is increased a little. During the treatment his temperature went up from 98·4° to 101·8° F. His pulse, which was 70 at the beginning, was 96 at the close.

CASE 6.

Traumatic Arthritis without Rheumatism or Gout.—This is a patient 70 years of age, kindly referred by Dr. Robert F. Weir, from his service at the New York Hospital. The history obtained is that he was thrown from a car about a year ago, his left knee being the joint injured. The pain has increased from the time of the injury. The stiffness also has increased. He claims that the treatment by fixation, cautery, and such other applications as have been made from time to time, have been ineffectual in arresting the progress of the disease. At all events, he is unable to walk, except by assistance. The range of motion is less than 10°. There is considerable tenderness on handling the limb. He is subjected to the Tallerman treatment, at a temperature of a little over 200° F., for fifty minutes, and on removing him from the cylinder, the house-surgeon, Dr. Kenerson, moves the limb, and is satisfied that an increase of 15° in motion has been gained.*

CASE 7.

Treated at the Children's Hospital, St. Michael's Hill, Bristol, at a demonstration.

W. A.; aged 47. Left knee painful and stiff after an accident two years ago. Can bend knee, but only to a small extent and with pain.

One bath; temperature 245°; duration forty-five minutes. After the bath patient could bend his knee without any difficulty. He said all pain was gone, and he could walk with perfect comfort.

E. C. WILLIAMS, M.B. Cantab.

CASE 8.

Treated at St. Bartholomew's Hospital. Under the care of MR. WILLETT, F.R.C.S.

Cecilia L., aged 17, was admitted into President Ward on January 16 suffering from painful knee, believed to be a subacute tubercular affection. In June, 1893, she had been in President Ward under Mr. Cripps, and the limb was then brought into a good position and put up in plaster of Paris. Some enlarged cervical glands at that time were also removed, and hence there is good reason for believing that the diagnosis of tubercular affection was correct. In the meantime she had been from time to time treated by reapplication of plaster of Paris splints, and once leather splints were moulded to her knee, and worn for a time. Recently there had been an exacerbation of pain and swelling, and for this she was readmitted. She was of the rather dull, sleepy type of patient. There was some backward displacement of the tibia, with a tendency to rotation of the tibia inwards rather than outwards. The patient usually lay, not on the affected side, but upon the opposite side, and rolled her limb inwards. This, of course, kept the crucial ligaments tight, and also prevented backward and outward displacement. On January 17, on account of localized pain, an icebag was applied; and in February a ten-pound weight was adjusted, position remaining good. On February 3 the limb was again put

* Read before, and demonstration given to, Practitioners' Society, New York, December 4, 1896. I am indebted to Mr. Lewis A. Tallerman, of London, for the opportunity of using the apparatus in the cases here reported.—V. P. G.

into plaster of Paris. It was taken out on the 9th. On the 20th the hot-air bath, at a temperature of 170° F., was used for half an hour. This did not seem to increase the movement, but did seem to diminish the pain. Ultimately she left the hospital wearing a Thomas's knee-splint. The leg was straight and firm and much less painful, and her general state of health was improved. This patient had the bath on only two or three occasions. Probably the affection was rather too acute, or, if not so, the patient was of that neurotic type in whom it is difficult to tell whether the pain is real or imagined. But it did not, on the whole, seem a very fair test case for the bath treatment; she complained so greatly of it that it was soon abandoned. I think she only had some three baths in about two weeks.

CASE 9.

Treated at the North-west London Hospital. Under the care of MR. MAYO COLLIER, F.R.C.S.

L. B.; aged 12. Tubercular knee-joint. Had been kept at absolute rest in plaster of Paris for upwards of seven months. On removing the splint the disease was found to have disappeared, leaving little or no movement, any attempt to increase it causing severe pain.

After six operations the movement had increased 45° to 50°; there was no grating in the joint, and the patient could walk freely and well; no forcible movement was used.

CASE 10.

Treated at the Livingstone Cottage Hospital, Dartford.

Acute Synovitis of Knee-joint.—G. L.; aged 63. Labourer. Strained his knee-joint three weeks ago when at work as a bricklayer; the joint quickly began to swell, and he has been incapacitated ever since.

I saw him for the first time on March 25, when I found the right knee-joint slightly distended with fluid; there was no increase of heat, but great pain was caused on manipulation or on attempting to stand.

The patient was treated by the superheated dry-air apparatus the following day, the right leg being kept in the cylinder for forty minutes, and exposed to a temperature of 250°. On removal, the joint could be manipulated with but little pain.

A second application was made on March 29. In the

interval the swelling of the knee-joint had almost entirely disappeared, and the pain had been very trivial. After this second application the joint could be moved quite freely, and without causing any pain.

Two days afterwards the patient was able to resume his employment.

A further point noticeable in this case was in regard to the feet. These had been operated upon in infancy for talipes—from all appearances equino-varus—and there had always existed a considerable amount of stiffness about the joints. This was relieved to a most marked extent by the hot-air bath applications.

September 5, 1896.—The improvement in this case continues.

CASE 11.

Treated at the Royal Portsmouth Hospital. Under the care of MR. D. WARD COUSINS, F.R.C.S.

Chronic Traumatic Synovitis. — J. L.; aged 49. Labourer. Chronic synovitis of knee-joint over two years. History of injury, by fall. Before the superheated dry-air treatment was commenced, other methods had been resorted to for several months without producing any satisfactory results. Knee considerably swollen; kept always in a position of considerable flexion; gave pain on movement; and also when at rest very limited range of movement. Disease had made considerable progress and affected the bones, in addition to the synovial membrane of the knee-joint. The treatment was continued for the whole of the month during which the apparatus was available, and in all the hot-air bath was applied nineteen times. On cessation of treatment the circumference of the limb, measured above the patella, over it and below it, had diminished by half an inch at each point. Range of movement also slightly increased. Pain undoubtedly considerably relieved.

T. H. BISHOP, M.B., C.M., House-Surgeon.
H. W. MORLEY, M.R.C.S., L.R.C.P., Assist. H.-S.

CASE 12.

Treated at St. Bartholomew's Hospital. Under the care of MR. W. J. WALSHAM, F.R.C.S.

T. W. Builder. Potts' fracture one year ago. Joint painful, and bones in bad position.

Osteotomy of tibia and fibula, with removal of wedge-shaped piece of tibia.

Position much improved; pain on walking still.

This patient complained greatly of the pain on moving ankle-joint.

He was treated by the Tallerman hot-air method; altogether he had six or seven baths. The movement in the ankle remained unaltered, but he expressed himself as much better as regards the pain. This improvement, he has since informed me, has continued.

MARTIN JONES, M.R.C.S., House-Surgeon.

CASE 13.

Treated at St. Bartholomew's Hospital. Under the care of MR. W. J. WALSHAM, F.R.C.S.

J. J.; aged 35. Fractured leg. Treated plaster of Paris.

Condition.—Joint stiff (ankle); painful on forced movement. He was treated twice or three times by the hot-air bath. He was much improved; movement gave him much less pain, and he walked with much greater comfort.

MARTIN JONES, M.R.C.S.

CASE 14.

Treated at the Royal Portsmouth Hospital. Under the care of MR. D. WARD COUSINS, F.R.C.S.

S. C.; aged 50. Fell downstairs six weeks ago, and sustained severe sprain of wrist; wore a splint for six weeks, and then consulted a doctor, who advised her to undergo treatment in hot-air cylinder. Came to hospital on October 8.

On examination, the wrist and fingers were found to be extended perfectly stiff, and attempts at passive motion caused great pain. Patient herself could not move any of the fingers, but could touch base of index-finger with the thumb. She was reluctant at having the adhesions first broken down under an anæsthetic, so that it was determined to try the effects of the Tallerman treatment without.

First Operation. October 9.

Hand and arm placed in cylinder for forty minutes at temperature of 240° F. After fifteen or twenty minutes she said she could move her forefinger slightly. On withdrawing the hand from the cylinder it was found that the fingers could be moved slightly without much pain, and she could herself touch the tip of her index-finger with her thumb. Although before the treatment she complained of great pain on attempting to flex the wrist, after the bath it was partially flexed, and several adhesions broke with an audible snap, the patient experiencing but little pain.

Second Operation, October 11. *Forty minutes at* 290° *F.*

Movement in fingers and wrist still greatly increased. Patient could touch the tips of all her fingers with her thumb. Some more adhesions in wrist were broken without much pain.

Third Operation, October 13. *Thirty minutes;* 280° *F.*

Movement in fingers and wrist markedly increased. She can now flex all her fingers into the palm, and has a good range of motion in wrist-joint.

Saw patient again on October 28. She can now make a fist, and move her wrist-joint freely. Is greatly pleased at the result of the treatment, and has resumed her household duties. Pain on movement is but trifling.

T. H. BISHOP, M.B., C.M., House-Surgeon.
H. W. MORLEY, M.R.C.S., L.R.C.P.

CASE 15.

Treated at the Royal Portsmouth Hospital. Under the care of MR. D. WARD COUSINS, F.R.C.S.

H. W.; aged 12. Stiff fingers after a fracture of forearm three months ago.

First Operation, October 3. *Forty minutes;* 265° *F.*
Considerable improvement; fingers much less stiff.

Second Operation, October 4. *Thirty minutes;* 240° *F.*

The stiffness of the fingers has greatly improved; can fully flex the fist, and fully extend all the fingers with the exception of the little and ring, which cannot be quite straightened.

Third Operation, October 9. *Thirty-five minutes;* 230° *F.*

All the fingers can be fully flexed and extended without pain. When last seen (November 2) his hand was perfectly well.

T. H. BISHOP, M.B., C.M., House-Surgeon.
H. W. MORLEY, M.R.C.S., L.R.C.P.

Case 16.

Treated at the Royal Portsmouth Hospital. Under the care of Mr. D. WARD COUSINS, F.R.C.S.

E. B.; aged 31. Stiff fingers after collis fracture nine weeks ago. All fingers are semi-flexed. Unable to straighten them herself, and attempts to forcibly extend them cause severe pain.

First Operation, October 4. Forty minutes; 250° F.

After the bath the fingers could be further extended with less pain; straight splint applied.

Operation repeated on October 9, 11, and 16; on each occasion for thirty minutes, at temperature of from 240° to 265° F. At the end of the treatment she could freely extend the fingers without pain, and she has now resumed her occupation, and is quite well.

T. H. BISHOP, M.B., C.M., House-Surgeon.
H. W. MORLEY, M.R.C.S., L.R.C.P.

Case 17.

Treated at Charing Cross Hospital. Under the care of MR. MORGAN, F.R.C.S.

W. D. Injury to hand. Under treatment twelve weeks. When the wound was healed there remained stiffness at wrist-joint, and also metacarpo-phalangeal and inter-phalangeal joints. Massage for some six or eight weeks improved matters to a degree, but it was thought advisable to adopt the superheated dry-air treatment by the Tallerman process. This was done for three weeks, with a marked improvement as result.

C. LOCKYER, M.R.C.S., L.R.C.P., House-Surgeon.

Case 18.

Treated at the Livingstone Cottage Hospital, Dartford.

Injury to Metatarso-Phalangeal Joint. — R. M.; aged 16. Domestic servant. About three months ago a heavy tea-tray fell on her left foot, causing severe pain over the metatarso-phalangeal joint of the second toe, which was considerably increased on movement or pressure. She was obliged to give up her situation in consequence.

The case came under my care a month after, when, finding the pain as acute as ever on the patient standing or bearing her weight on the foot, I put the foot up in a plaster case. This was continued for a fortnight, but soon after removing the case the painful condition returned. Painting iodine over the joint had no effect.

On April 12 the patient's foot was placed in the hot-air bath for forty-five minutes, the temperature of the bath reaching 240°.

On removal she could bear her full weight on the foot without pain, and pressure over the joint and flexion of the joint caused no pain.

I saw the patient again a fortnight later, and found the pain had not returned, and she was anxious to again go out to service.

August 20, 1896.—There has been no return of the pain, movement of the foot being quite free and causing not the slightest discomfort.

CASE 19.

Treated at the same Hospital as Preceding.

Stiffness after Fractured Humerus.—A. C., aged 16, fractured his right humerus four weeks ago, since when it had been kept in splints.

On removing the splints there was slight stiffness of the elbow-joint, and considerable stiffness at the shoulder, the slightest movement causing great pain.

The right arm was placed in the hot-air bath on April 13, four weeks and five days after the injury.

The application was continued for about forty minutes, and the temperature of the bath was raised to about 260°. Immediately afterwards the limb could be moved freely both at shoulder and elbow without causing any pain, and the patient has had full and painless use of that arm ever since.

September 5, 1896.—This patient has had full use of the right arm ever since, and without pain.

CASE 20.

Treated at the Liverpool Infirmary.

Periostitis.—John G., aged 43, admitted to hospital suffering from periostitis around ankles, October 26, 1896. Boracic fomentations and McIntyre's splints applied, but no relief experienced. Tallerman baths commenced November 14 and continued till 28th. Had eleven baths. Patient's highest temperature in bath 99·4°; highest temperature of bath 250°. Went to town December 1, quite well, and able to walk; no pain.

CHAPTER VI.

GOUT.

EXCELLENT results have been obtained, as will be seen from the following eighteen cases, both in acute and chronic gout. In acute attacks inflammation and pain are reduced with extraordinary rapidity, while in chronic gouty conditions the morbid products in the blood are effectually eliminated by the skin and the kidneys. On this head we cannot do better than quote the observations made by Dr. Sibley in a paper published in the *Lancet* on July 10, 1897. His remarks, together with a series of ten cases, are here produced. The other cases include an interesting one of saturnine gout, treated at the Philadelphia Hospital:

'A few remarks are necessary with regard to the mode of action of this treatment. The heat applied locally to the inflamed gouty part causes an increased circulation in the area, bringing a larger quantity of blood to the seat of the lesion, and at the same time taking away a larger quantity of blood, probably more or less saturated with the bodies causing the local deposits. These deleterious products being in the general circulation, so to speak, dissolved out of the seat of the inflammation, should in the normal course of events largely find their way out of the body through the kidneys; hence all should be done during the course of treatment to assist these organs to act as freely as possible. Some interesting observations on this part of the subject have recently been made in the clinics of Professor Landouzy, at the Laennec Hospital, and Dr. Déjerine, at the Salpêtrière Hospital, in Paris, who are experimenting with the Tallerman treatment, the results of which were published by Dr. Chrétien.* They found, for instance, that in one case of ordinary gout the daily elimination of uric acid, which after the fourth bath was 57 centigrammes, rose after the ninth bath to 89 centigrammes; and in another case of arthritis the daily co-efficient of urea had changed from 20 grammes 97 centigrammes before treatment to 25 grammes 50 centigrammes after the treatment had been administered. They also found an increased excretion of all the salts, especially the chlorides. This increased excretion through the kidneys is probably the explanation of the great benefit of the hot-air treatment in cases of gout. The results in the French hospitals were obtained entirely by this external method, and without the exhibition of drugs of any kind. I feel no doubt of the great benefit this treatment will show in cases of what are described as "constitutional goutiness," without necessarily any local objective lesion.'—*Lancet,* July 10, 1897.

CASES 1-10.

Treated at the North-west London Hospital. Under the care of DR. KNOWSLEY SIBLEY.

CASE 1.

Acute Gout.—The first case was a man, aged 65 years, who came under observation on June 3, 1896. The patient had suffered from acute attacks of gout on and off for many years, the attacks often lasting for several months. A year and a half previously he had had a bad attack of pneumonia, followed by a prolonged attack of gout in both feet and one knee, which lasted for three months. The present attack had commenced a few days before in the left elbow, which became swollen and the skin tense, red, and very painful; the back of the hand was also swollen, the knuckles being hardly visible, and the veins much engorged. After the first treatment by the superheated dry-air bath, the parts about the elbow were not so tense, there was less œdema over the back of the hand, and the veins were not so distended. The patient had a return of the attack in the evening in the elbow, accompanied by severe pain; after

* *La Presse Médicale,* December 26, 1896.

treatment on the following day the elbow appeared less inflamed, and could be moved better; the pain was also greatly relieved. On the evening of June 5 the arm again became more painful, but the patient had a good night; the following day there was some effusion into the tissues below the elbow. Treatment again produced considerable improvement of the parts. By June 8 the gout had quite subsided in the elbow; there was no pain or effusion left, and the joint could be freely moved. During the night the index-finger became inflamed; this became more comfortable after treatment, and the swelling rapidly subsided. The patient went into the country for a change. In April, 1897, the patient had remained quite well, and had been free from attacks of gout since the treatment. During the treatment he was taking alkaline medicine and a saline aperient.

CASE 2.

Acute, Subacute, and Chronic Gout associated with Eczema.—This patient was a lady, aged 50 years, who commenced this treatment on June 25, 1896. There was no history of gout or rhematism in the family; her father lived to the age of 90 years, and her mother lived 73 years, and suffered from eczema. Her paternal great-grandfather lived to the age of 105 years. The patient when young had excellent health till the first attack of gout, twenty-four years ago, in the right great toe, when she was laid up for a fortnight. From this time she had recurrent attacks of acute gout every three or four months. For many years it was confined to the feet. It first appeared in the knees twelve years previously, and in the hands for the first time seven years previously. The attacks had been much worse the last six years; she was often laid up in bed for from six to ten weeks, and in acute pain. Five years previously she had an attack of eczema in the feet, and this had troubled her on and off ever since, never having disappeared. Eight years ago the chalky, gouty deposit in the right great toe broke down and discharged for twelve months, and this place had reopened on and off several times since. In October, 1895, the deposit in the left index-finger discharged for four months, and about the same time the left great toe also discharged. From October, 1895, to June 25, 1896, the patient had been laid up in bed, being unable to move because of pain, although taking large quantities of medicine. The left index-finger was much deformed, and chalky deposit was seen beneath the thinned skin. The little finger of the right hand was also much deformed; there was chalk-stone deposit in the middle finger of the left hand. The left knee was extremely painful and tender, movement was very limited on account of pain, and there were some deposits to be seen and felt on the anterior surface of the patella. Both the feet were swollen and covered with an irritable form of eczema, which, especially on the left, extended some way up the leg; both great toes were much deformed, tumid, and inflamed, very red, and acutely painful; they were both dislocated outwards, and the left was discharging thick pus and chalky matter, and looked very angry. There was considerable puffiness about the ankles, and pain on the slightest attempt at movement. There were deposits in both ears. On June 25 the left leg was placed in the Tallerman apparatus. The pain was soon relieved by the heat, and when taken out she was able to move the limb freely, and could even walk on it without pain; the eczematous condition was also better. The next day she reported that she had slept all night, that the knee was almost well and the foot much less painful, and that there had been no irritation from the eczema, although for the first time for four months she had not used any ointment. She had a second bath, and was again much relieved. She returned for a third bath on July 6, when she reported that she had had another attack of eczema, which was very distressing. After treatment the intense irritation was much relieved; the toe continued to discharge freely. The patient's foot was too bad to allow her to come for treatment from July 8 to July 22, when, after a bath, the part was much better, and she felt very comfortable. By August 20 the patient had had ten baths. The fourth finger was now decidedly smaller, and the chalky deposit in the knuckle of the second finger was dispersing. On the 25th the patient had her eleventh bath, and said she was able to play the piano for the first time for six months. On the 26th, after the twelfth operation, the following note was made: 'The various chalky deposits are decreasing quickly and becoming much softer; the fingers are more flexible, and the whole hands are of a much better shape.' On September 2 the patient informed me that she had been trying her voice, and had found that it was quite restored; she had not sung for eighteen months. She was also able to run up and down stairs; the hands continued to improve. On the 29th the patient had her twentieth bath. She had had a bad cold, and on the previous day, after doing some needlework, she felt a sudden sharp pain from the right shoulder to the tip of the second finger on the right hand, and this had become stiff and painful. Now the elbow was also painful, the tissues about it being tense and swollen, and tender to the touch. All these symptoms were much relieved by another application of the hot air. The next day she reported having had a good night, and there had been no pain in the finger or elbow since; she had been able to dress herself. On December 16 the patient again presented herself, not having attended since September, owing to numerous domestic worries. She had been well, with the exception of an attack of pain under the knees and down the calves (? phlebitis) in November. At this time the patient was fairly well in herself, but had some gout about the ankles and heels; the eczema had quite disappeared from the feet. The right elbow was enlarged, red, and painful. On the 24th the swelling about the left elbow which was present last time had increased, and had become extremely painful. On the 29th the patient reported that her left elbow was much less painful for two days after the last bath, but it had then again become worse; the right elbow had not since been painful. On the 30th the left elbow was very red, swollen, and painful over the olecranon process; there was also a good deal of pain up the arm (lymphatic), and also in this axilla. The pain and congestion of the whole was much relieved by the bath. The next day the patient reported that the elbow again became painful during the evening, but she had had a good night and was feeling much better. During the next few days the left elbow kept very painful, and the skin broke on January 5, 1897, and it then discharged slightly. A day or two afterwards the right arm became painful, then the left became still more inflamed, and the patient had to keep in bed. She had had no sleep for several nights, so she was taken in at the Tallerman Treatment Institute for treatment on January 13. The left arm, which was greatly inflamed and red from the elbow to the middle of the arm, was placed in the cylinder, after which the patient fell asleep. Having slept for three hours, she awoke feeling much more comfortable. She did not have a good night, the elbow discharged very freely a thick chalky material, but in the morning the swelling and inflammation had considerably subsided. The whole condition rapidly improved during the next few

CASE 2.—Before treatment: showing enlargement of great toe and eczema.

CASE 2.—Taken March 30, 1897; showing marked improvement in condition of left great toe and of hands.

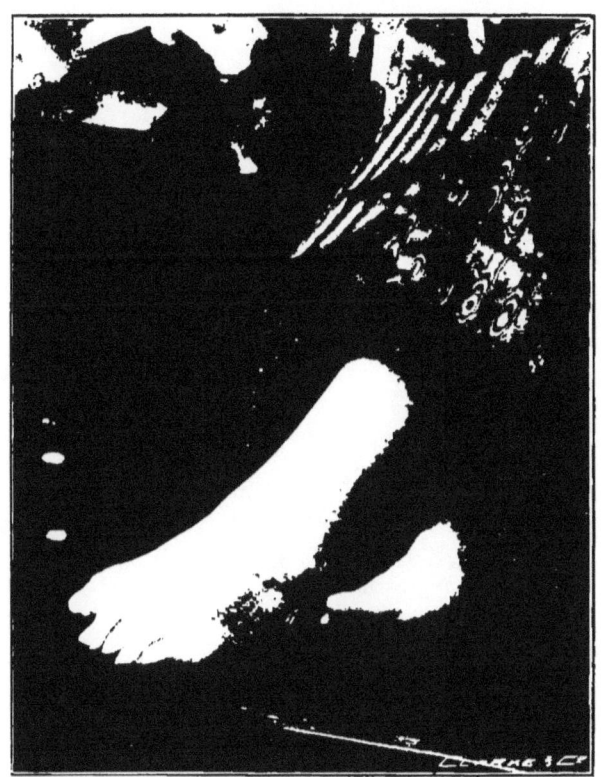

CASE 2.—After treatment: showing absence of any eczema or gout.

To face p. 92 –IV.

days. By January 21 the patient was well, and she returned home. The elbow had ceased to discharge, and the movements of the joints were unimpaired. She stated that the last time her finger discharged it continued to do so for six months. In April, 1897, the patient had continued quite free from gout since January, and there had been no return of the eczema; her general health had also greatly improved.

CASE 3.

Subacute Gout, Chronic Bronchitis, and Albuminuria.— This patient, who was first seen on September 21, 1896, had probably inherited gout from his mother's side of the family. A painter by trade, he had had severe colic fourteen years ago, and was 61 years of age. The first attack of gout occurred ten years ago in the right foot, when he was laid up for a fortnight; a second attack came on three years later, and for the last three years he had had repeated attacks, especially in the ankles. When he came to the hospital the right foot was swollen, very tender, and painful, the skin being red and much inflamed; both legs were œdematous. Internal remedies were prescribed. On October 12 the general condition was much the same. The cough and bronchitis continued troublesome, with a good deal of expectoration in the early morning. There were general bronchial rhonchi, especially at the right apex. The urine was clear, of specific gravity 1012, and contained albumin. The second heart sound was accentuated. The condition of the foot was much about the same, the patient being hardly able to get about, and then only with great pain. On the 13th the first hot-air application was given, no change being made in the medicines. The foot was much easier, and he could move it more fully after treatment. By the 19th he had had three baths. The foot was much better, and the bronchitis was also much improved; there appeared to be less albumin than formerly. He had slept much better since the baths were administered. On the 20th the patient stated that his foot felt quite well; he had no pain whatever. He took a mixture of gentian and bicarbonate of soda all through the treatment.

CASE 4.

*Acute Gout.—*This case of acute gout in a man, affecting the left ankle, was greatly relieved by three baths.

Case 5.

Acute Gout.—A man, aged 43 years, was first seen on November 13, 1896. His father died from rheumatism and Bright's disease. Gout occurred for the first time ten years ago, and the attacks had been repeated at least once a year ever since. This attack came on two days previously with great pain, chiefly in the left foot. He was given an alkaline mixture of soda and gentian and the hot-air baths. The patient felt great relief from the first bath; he could bend the foot with comfort afterwards, and the pain was much less. On the 15th the foot had been much better. There was no pain, but some stiffness about the ankle when walking. After another bath the foot felt quite well, and the stiffness had disappeared and did not return. On April 3, 1897, the patient wrote to say that he had not had a return of gout in the feet, and had continued at work ever since the previous November.

Case 6.

Acute Gout.—A man, 24 years of age, came under observation on December 7, 1896. There was no family history of gout or rheumatism. Fourteen weeks previously the patient had suffered from inflammation of the eyes, which lasted three weeks, followed by pain in the back, which kept him in bed for three weeks; afterwards the pain settled in the knees, the left great toe, and then in the left hand and arm. Between December 7 and 21 he had four baths, and in addition a mixture containing bicarbonate of soda and iodide of potassium. By January 4, 1897, he was quite free from gout.

Case 7.

Gout, Bronchitis, Heart Disease, Albuminuria. — This patient, who was a woman aged 65 years, came under observation on November 23, 1896. Her maternal grandmother, who suffered from gout, died when over eighty, and her mother died at 93 years of age. She was quite well till 43 years of age, when she had the first attack of gout in the left great toe. She had had frequent attacks since in the feet, ankles, knees, elbows and hands. For the last eighteen years she had also suffered much from bronchitis. On December 8 she was admitted into the hospital. At this time the patient was very weak and suffered much from dyspnœa; the pulse was rapid (120) and small; the cardiac dulness was increased; there was a soft systolic murmur at the apex and a much louder one at the base, which murmur was readily heard in the vessels of the neck; the heart sounds were very feeble; there were general bronchial rhonchi over both the lungs. The urine was of specific gravity 1010, acid and clear, and there was some albumin present. The feet were much distorted from old standing gouty changes; there was a large nodule in the left metatarsal phalangeal articulation of the great toe. There was very little movement in the knee-joints, and there was thickening about the heads of the tibiæ. The joints of all the fingers were enlarged, and the fingers deflected to the ulnar side. There was a large firm nodule over the right olecranon, and hard nodules on the outer side of the olecranon of the left elbow. The patient had had little or no sleep for weeks, and the breathing was rapid and shallow. On December 11, after the first bath, the patient seemed to be a little easier. By the 14th the breathing was better after the bath, but she had not slept much. By the 21st the right hand was not so painful, and the cough was a little easier; the patient still felt very weak, and her skin did not act well. She gradually improved, and left the hospital on January 7, 1897.

CASE 8.

Subacute Gout.—This patient, who was a cab-driver, 38 years of age, was first seen on January 4, 1897. He was treated in the same way as the patient in Case 6, being given two baths, and by January 14 he was free from pain.

CASE 9.

Acute Gout.—This patient, a woman, aged 62 years, came to the hospital on February 22, 1897, and gave no family history of gout or rheumatism. She had congenital absence of the first, second and third fingers of the left hand. Eighteen years ago she had her first attack of gout in the right great toe, and she had had repeated attacks, often twice a year, ever since. The present attack came on a week ago in the left shoulder, then the right leg, and three days later the right hand became involved. The whole hand was much swollen, very red, and the skin much inflamed and acutely painful; there was some inflammation of the left ankle. She also suffered from chronic bronchitis. The first application of the Tallerman treatment was administered on the 23rd, with the result that the hand was less swollen and felt much easier after the bath. A second was given on the 25th. The fingers continued to be very stiff, but there had been much less pain in them. A third application was given on March 1, when the patient said that she had slept much better and the pain had been much relieved. After this she considered herself cured, and returned to her occupation on March 10.

CASE 10.

Acute Gout.—This patient was a salesman, who came to the hospital on March 21, 1897, suffering from acute gout in both hands, which was quite relieved by three baths and an alkaline mixture.

It is important to note that nearly all of the above cases were treated as hospital out-patients, and that therefore it was impossible to regulate or in any way control their daily habits as to diet, etc. Most of the patients were also only treated during an attack of gout. It was difficult to persuade them to undergo treatment by the hot air between the attacks, even with a fair expectation of keeping off a return of the disease. For instance, in Case 2, although many promises were made, something always interfered to prevent the continuance of regular treatment. In spite of the considerable shortening of the attacks and of the improvement in her general condition, freely admitted by the patient, only the return of pain would persuade her of the need to continue with the hot-air applications. My experience with these cases is that directly the pain subsides the patient returns to work, and neglects to take medicine or the ordinary diet precautions. The cases above described contain examples of gout in its various stages, from the most acute, through the subacute, to the very chronic, these latter, sometimes with acute exacerbations, occurring from time to time. Taking first the very acute attacks (Cases 1, 4, 5, 6 and 9), in all of these the local intense pain and congestion were almost immediately relieved by the hot air; very obvious difference in the appearance of the part was seen after the application, and the relief from the intense agony of an acutely inflamed gouty joint was most marked. But gout, like rheumatism, is always prone to reappear, and it was found in Cases 1 and 6 that the acute inflammation in the part first affected rapidly improved under the local treatment, and in the course of a few days another deposit, but one not necessarily so severe, appeared in another part, sometimes the corresponding limb of the opposite side. Treatment would again rapidly disperse this second deposit, and in no instance did it afterwards follow in another joint. It would seem that in nearly all the cases the attack under the hot-air treatment was very much shorter than the previous attacks which had been treated only by internal remedies (Cases 1, 2, 4 and 5). Often, in fact, the attack which previously had lasted for weeks under this method lasted only for the same number of days. In some (Cases 2, 6, 8 and 9) internal medicines, such as small doses of colchicum and iodide of potassium, were also given; in others only a simple alkaline mixture (with a slight aperient) was administered, as in Cases 1, 3, 5 and 10. Nevertheless, in all of these cases the relief from pain and the rapid and complete disappearance of the gout were most marked. Case 2 very well illustrates the treatment of very chronic gout, with gouty or chalk-stone deposits; the deposit in one finger for instance, being so bad that at one time the medical attendant had wished to amputate the digit, and yet after several applications of heat the deposit gradually disappeared, and so also the deposits about the heads of the larger bones became distinctly lessened.

Case 11.

Treated at the Laennec Hospital, Paris. Under the care of Professor Landouzy.

Acute Gouty Arthritis (?) of Palms and Surrounding Joints (Arthrite Radio-Palmaire aigue goutteuse ?). — Maurice G.; 24 years old. A medical student. He gave a history of hereditary arthritis (rheumatism, gout, etc.). He had, when 16 years old, a severe attack in the toes of the left foot. He appeared in the hospital with a red swelling, inflamed and painful in the right wrist and hand, which he could not use at all. The fingers were inseparable and bent, the arm held in a sling.

After ten minutes in the apparatus the patient began to feel his fingers move; at the end of the treatment, which lasted about fifty minutes, the pain had completely disappeared; the swelling was diminished, the movements easier. Three hours afterwards the patient played the piano without difficulty, and the next day he covered thirty kilometres on his bicycle. Three days afterwards he had a slight attack of pain in the left hand, which disappeared completely after one treatment.

Case 12.

Treated at the Philadelphia Hospital.

Saturnine Gout.—Thomas Latham ; aged 46 ; birthplace, England ; residence, Philadelphia ; occupation, worker in lead ; attending physician, F. A. Packard ; resident physician, Raymond Spear.

Mother died of asthma. Father died of old age. One daughter living. Three brothers and two sisters living and healthy. Has always been healthy. No venereal history. No history of alcoholism. Has worked in lead for many years. Since 1890 has had numerous attacks of gout, which has affected the small joints in his feet and hands. For past two years has been an invalid. Had suffered with pains all through his body. Has numerous tophi, situated in small bones of hands and feet. His heart is much enlarged. His arteries show evidences of arteris and capillary fibrosis. His urine contains albumin, at times one-third bulk by boiling test, and some granular and hyaline casts. On November 11, 1896, he was taken to the College of Physicians and Surgeons and exhibited before the County Medical Society. When exhibited he had an acute attack of gout, involving point of great toe of right foot. He had shooting pains through his body, a severe headache, his grip was weak, and the joints of his feet, especially his right, were immobile. He was placed in a Tallerman Hot-Air Bath for forty minutes. After the bath the reddened, angry joint of his great toe appeared white. The swelling had disappeared. The joints of both feet were freely movable. The pain had disappeared. His grip was much stronger. His headache had left him. He said he had not felt so well since he had been taken sick. On November 11 the man complained of headache and pains in his left leg (the foot that had not been treated), also shooting pains through the rest of his body. Was given another Tallerman Hot-Air Bath. Before the bath the joint of the great toe of his left foot was the seat of an acute attack of gout. After the bath the pain had disappeared, and the joint resumed a normal appearance. The pains in the rest of his body had disappeared. The results of the treatment in this case were remarkable, owing to the rapidity in the disappearance of symptoms which until this time had resisted treatment for the past two years.

CASE 13.

Under the care of ARTHUR ROBERTS, M.D.

Acute Gout.—A. H., aged 51 years, was suffering from acute gout in both knees and feet, and left arm.

Had suffered for twenty years from gout, but up to now had always been able to walk about on crutches. Had been treated by medical men at various places, but received no permanent benefit, nor had he ever at any time been entirely free from the stiffness in the joints, though he had often been better and worse.

Now he is completely laid up, and for several nights has not been able to sleep owing to great pain.

I considered the case a suitable one for treatment by the Tallerman Hot-Air Bath mentioned by Mr. Willett in his clinical lecture at St. Bartholomew's; and a cylinder was obtained from London.

On June 4, 1895, the patient was treated for the first time.

Before treatment: Temp. 99°, P. 80, R. 20.

The left leg was put in the cylinder. Twenty-five minutes later he could straighten the leg and bear pressure on the knee without any pain. After the operation all the pain was gone, both in left leg and also in the other leg and arm.

Operation, forty-five minutes. Temp. after bath 99°, P. 84, R. 20.

Measurement of radial artery by arteriometer:

Before, 1·8 mm.
During, 1·9 mm.
After, 2·4 mm.

All these observations were taken in the recumbent position.

Patient said he 'felt very much better, and the bath was worth all the expense and trouble.'

Second treatment, June 5, 1895. Left leg in forty-five minutes. Patient finds all his pain gone. The pain had not returned since last treated.

Arteriometer:

After twenty minutes, 1·9 mm.
After bath, 2·5 mm.

Third treatment, June 6, 1895. Pain has not returned. Right leg was placed in the cylinder.

Arteriometer:

Before { Right, 2·8 mm.
        Left, 3·5 mm.
After { Right, 3·5 mm.
        Left, 4·0 mm.

This showed that the effect of the bath was to increase the calibre of the radial artery, and the improvement was not only in the leg, but also in the general system.

Fourth treatment, June 8. The left arm was placed in. I was not present on this occasion.

Arteriometer before bath:

Right radial, 2·5 mm.

I saw patient after the bath, and he reports continued improvement. Patient says that he has never had any treatment that has produced such a profound and beneficial effect on his system before. He is delighted with it.

The duration of the operations averaged about forty-five minutes, and the temperature 240° F.

Case still under treatment (June 8, 1895).

ARTHUR ROBERTS, M.D.

CASE 14.

Treated at the North-west London Hospital. Under the care of MR. MAYO COLLIER, F.R.C.S.

W. E. E.; aged 57 years. Goods guard, L. and N.W.R. Patient sprained his right ankle whilst at work on August 11, by striking it against a stack of iron.

On the day after the injury he had a typical attack of gout in the great toe of the injured foot, and stated that he had been unable to sleep for two nights, and was in great pain.

On examination the right ankle and foot were very much swollen, the joint filled with fluid, and the entire surface reddened and shiny. Acute tenderness was present over the whole foot, especially over the tarso-metatarsal joint of the great toe; all movements were painful, and the foot could not be put to the ground.

First Operation, August 14. *Forty minutes.*

After the bath, the pain and tenderness to pressure and movement had gone from the ankle-joint, and the pain and tenderness in the great toe were much modified.

Second Operation, August 15. Thirty minutes.

Patient had passed a good night, and after the operation he was able to stand.

Third Operation, August 20. Thirty minutes.

On examination there was no pain or tenderness on movement or pressure, and the patient was now able to walk. Some redness still remained.

Fourth Operation, August 21. Forty minutes.

Swelling almost gone; no pain or tenderness.

Fifth Operation, August 23. Forty-five minutes.

The patient having reported that he had been able to walk for twenty minutes without pain, and all effusion having disappeared, the ankle and great toe being now freely movable and without pain, he was considered in a fit state to resume his ordinary duties, and was discharged.

Note.—The temperature in this case was maintained at from 230° to 240° F., and the result obtained was most satisfactory. The pain was relieved almost at once, the effusion rapidly disappeared, and the patient was cured of a bad attack of gout with a severe sprain in twelve days.

J. F. SARGEANT, M.R.C.S., M.R.C.P.

CASE 15.

L. N., a gentleman, aged 53, who for some years past has been subject to occasional attacks of gout, had in April last a sharp seizure in one ankle. After the first bath of one hour the pain was gone, and he was able to put his foot to the ground. The treatment was repeated on the two following days, and he then returned to business.

B. MEREDITH ROWE, M.R.C.S.

Up to this date there has been no return, or other attack.

CASE 16.

Treated in South Charitable Infirmary, Cork.

Mrs. M. K. Gout in great toe; synovitis of ankle two months' duration. Had to give up work and was confined to bed. Unable to sleep. After first bath was able to sleep; after second walked to infirmary; after eight baths was completely well.

' I have much pleasure in stating that Mrs. M. K., suffering from gout and chronic inflammation of ankle-joint, was completely cured by Tallerman Hot-Air treatment.'

JOHN REID, M.B., B.Ch., Senior House-Surgeon.

Case 17.

Treated at the Laennec Hospital, Paris. Under the care of
Professor Landouzy.

Gout.—X.; 40 years old. Arthritic. Several attacks of genuine gout. Has tried all methods of treatment and has obtained only incomplete and passing results.

Persistent pain in the great toe and in the metatarsal joints of the right foot. Walking inconvenient, painful and often impossible. Swelling and hypertrophy of the affected joints. Measurements of the affected parts before the treatment: Right knee, 37 centimetres; left knee, 41 centimetres; right instep, 27 centimetres; left instep, 25 centimetres.

Treatment with the Tallerman apparatus.

Disappearance of pain at the first bath. It returned a little, only to disappear permanently after eleven baths. Movement restored. Swelling of the left knee and right instep reduced. Measurements after the sixth bath: Right knee, $36\frac{1}{2}$ centimetres; left knee, 39 centimetres; right instep, 25 centimetres; left instep, $24\frac{1}{2}$ centimetres. Thus there was a reduction of 2 centimetres in the left knee and of 2 centimetres in the right instep. The patient is still under treatment.

Case 18.

A. R.; aged 60. Cotton spinner. Suffering from acute gout in both feet, toes and ankles, both knees, and index-finger of right hand. All these joints are swollen, tense with effusion, exquisitely painful and immovable. The present attack has lasted about a fortnight, during which time he has had little or no sleep on account of the pain.

History.—Thirteen years ago suffered from shock, the result of a railway accident, and four days afterwards had his first attack of gout. Began with pains in the head, followed in a few days with pain and swelling in the ball of the great toe of the right foot. From this time till present date has had periodic attacks once or twice a year.

November 21.—In spite of treatment, the pain was so severe that I asked Mr. Tallerman, who very kindly consented, to treat the patient in the Tallerman Superheated Dry-Air Bath, with the following remarkable result:

Before the operation the patient lay helpless upon his back, having a drawn and anxious expression, absolutely unable to move, except his left arm, and his right arm only from the shoulder. Temperature 100·6°, and pulse 92 full.

Case 17.—Before treatment: showing swelling of left knee and right great toe.

To face p. 102--I.

17.—After treatment: showing reduction of swelling of left knee and right great toe.

To face p. 102—II.

The left leg was put into the cylinder, and the temperature raised from 150° to 220°, whilst the rest of the body was enveloped in a blanket. In ten minutes he was asked, 'Have you any pain?' when he thought for a moment, and said, 'No; the heat seems to have absorbed the whole of the pain.' After thirty minutes he was able, slowly and without pain, to move the fingers of his right hand so as to make a loose fist. At the end of forty minutes he was taken out of the cylinder. During the operation he perspired profusely, and his pulse had risen to 108, whilst the temperature was up to 101·6°. Twenty minutes later temperature had fallen to 100° and pulse 92. He was able to move both his ankles slowly, and had some degree of flexion in both knees, which, however, were still very tense. In about two hours the pain returned, and was very severe for about twenty minutes, when it disappeared, and he slept eight hours with one break of about ten minutes.

A second bath was given the following day, when it was noted that the temperature rose from 100° to 100·6°, and the pulse from 96 to 104. This bath was given for forty-five minutes, during which the temperature was raised from 160° to 220°.

November 26.—Patient convalescent; has had no return of the pain; able to be up in his room for six or seven hours.

December 10.—Right knee painful and stiff, temperature 99·6° and pulse 88.

December 14.—All pain gone. Patient downstairs. Pulse 80, temperature 98·2°; able to walk up and down stairs without inconvenience, though slowly.

Only two baths could be given, as Mr. Tallerman had to return to town and could not leave the cylinder, otherwise I feel sure the convalescence would have been much shortened.

Jos. G. G. Corkhill, M.B., etc.

CHAPTER VII.

SCIATICA, LUMBAGO, LOCAL PARALYSIS, WASTING, WRITER'S CRAMP, AND CHOREA.

The striking effects of the hot-air treatment in reducing pain have already been sufficiently commented on. Sciatica and similar affections therefore suggest themselves as particularly suitable for treatment, and the results obtained fully bear out that expectation. Some very severe and obstinate cases are here shown to have yielded readily, after other remedies had totally failed. As Dr. Sibley says: 'Very intractable cases of sciatica usually pass into the hands of the surgeon, who performs nerve stretching, often with very little good result. It seems likely that these cases will likewise in future be cured by this less drastic means.'

The two cases of partial paralysis included in this chapter are examples of another kind of nerve affection, which offers an extremely promising field. It is a pity that no more notes are as yet available, but other cases of paralysis and wasting of limbs have already been treated with the most happy results, and there is obviously scope for a wide extension of usefulness in this direction. Particular attention is drawn to the case of writer's cramp (Case 13). Patients' letters (of which a great many have been received) are excluded from this book; but an exception has been made in favour of Mrs. Walford, the distinguished authoress, who has been allowed to tell her own story, because of the literary interest attaching to her account. The accuracy of her description is vouched for by the medical man under whose care the treatment was conducted; its raciness and vigour speak for themselves.

CASE 1.
Treated at the Laennec Hospital. Under the care of PROFESSOR LANDOUZY.

Sciatica.—Louis D.; 32 years old. True sciatica had begun two months before, suddenly, on rising in the morning, and had resisted all therapeutic methods (chloride of methyl, blisters, salicylate of soda, sulphur baths, iodine). Unable to walk. Standing even very painful. Severe pains. Lasegue's symptoms.

Three Superheated Dry-Air baths; complete disappearance of pain. The patient left the hospital cured.

CASE 2.
From the same Clinique.

Sciatica.—Louis C.; 38 years old. The illness began in 1890, following a severe injury of the outside of the left thigh and hip. The pain returned in April, 1896.

Extremely severe pains along the great sciatic nerve, with radiations into the ankle and heel. Pain on pressure on the sacro-iliac joint. No new growth in the abdominal region was revealed by pressure in the region of the rectum and atrophy of the left leg. Various treatments tried without result. Salicylate of soda in large doses, siphonage, injections of chloroform. The patient took five baths. Under their influence the pain was much lessened, but did not disappear entirely. The patient is, however, able to walk, although limping a little. The sacro-iliac joint remains painful to the touch. Slight return of pain at the end of fifteen days. Result incomplete, as was to be expected from the undeterminate nature and rather peculiar character of sciatica.*

CASE 3.
Treated before the Practitioners' Society of New York. By V. P. GIBNEY, M.D.

Chronic Sciatica, with Recurring Acute Attacks; very obstinate.—The patient is a man, 50 years of age, who was referred by Dr. John J. Reid for this treatment. He had an attack fourteen years ago, lasting three months. The present attack has lasted four months. The pain is referred

* *La Presse Médicale*, December 26, 1896.

down the course of the limb, which is tender on pressure. He walks as if the limb were deformed, but the range of motion is quite free. He complains a great deal if he stands with the limb or walks with it. He is subjected to the bath this evening, during the meeting of the society, and, at the expiration of fifty minutes, temperature 260° F., he expresses himself as feeling much better and able to use the limb more freely. The immediate effect, therefore, is good.

V. P. GIBNEY, M.D.

CASE 4.

Treated by DR. A. ROBERTS.

Nurse Y. This was an old case of lumbago and sciatica with swollen knee. The right leg was treated for forty minutes at a temperature up to 240° F.

Result.—The patient was much improved and the pain manifestly lessened.

CASES 5-8.*

Treated at the North-west London Hospital. Under the care of DR. SIBLEY.

CASE 5.

Sciatica: duration seven months.—The patient was a man aged 56 years, who came under observation on July 16, 1896. His father, who died aged 75 years, suffered from rheumatic gout. The patient's brother also suffered from rheumatism. The patient had had rheumatic gout in his toes when 20 years of age, and had had muscular rheumatism occasionally ever since, but never very severely. He was in St. Thomas's Hospital for thirteen weeks in 1887 with nephritis and dropsy, and afterwards was an out-patient. For the last seven months he had had sciatica in the right leg, which had prevented him following his occupation. He was under a private practitioner for five weeks, and then became an in-patient in University College Hospital for a fortnight, but derived very little benefit; he had been attending as an out-patient till three weeks previously, but had gradually become worse. There was no albumin in the urine. He could walk a little, but with great pain and difficulty, and was able to raise the right leg but a few inches off the ground; there was much tenderness over the course of the sciatic nerve on pressure. On the 16th the right leg and part of the thigh were immersed in the cylinder for an hour at a temperature of

* From the *Lançet*, August 29, 1896.

CASE 5.—Sciatica (taken July 16, 1896), before treatment: showing limit of patient's ability to move. This is the highest point to which the limb could be raised, and the movement was accompanied by pain.

CASE 5.—Sciatic (taken July 20, 1896): showing patient's ability to move the leg after the third bath.

240° F. He was photographed with the leg raised as high as possible before and after treatment. The pain was much relieved, and he slept well that night, which he had not done for weeks. On the 17th the patient walked nearly five miles for treatment without much difficulty, and the hip was placed in the apparatus for forty-five minutes at a temperature of 250°, and the patient felt much better afterwards. On the 20th, after the third operation, he reported that there was a little pain above the right knee; the rest of the pain had quite gone; he could kneel, and he slept well. The treatment was stopped. On the 28th the patient had had no more pain; he had walked without a stick for the first time for six months, and he could flex the right thigh as high as the left. There was some stiffness about the knee, probably from having walked with a stiff joint for so long.

CASE 6.

Sciatica.—On July 27, 1896, a man aged 51 years came under treatment. He was a native of the West Indies. He had attended the Royal Free Hospital for debility a year previously, and had suffered from pains in the back and thighs since. Fourteen days before he had had a sharp attack of sciatica on the right side, and after a week he was compelled to give up work. He rubbed the part with embrocation, which had taken the skin off without giving any relief from the pain. An alkaline mixture and aperient pills were prescribed. On the 28th, the pain being no better, the patient was put into the Tallerman pelvic apparatus, after which the pain was much less. By the 30th he had had two baths; all the pain had gone from the hips and thigh, but there was still some pain in the front of the right tibia. On the 31st there was no return of pain in the thigh, but some in the lower part of the right leg. Another bath was given, and this time the right leg was put into the cylinder, after which the pain quite disappeared.

CASE 7.

Lumbago and Sciatica: duration six weeks.—On July 9, 1896, a man aged 26 years came under observation. There was no family history of rheumatism, but there was phthisis on his father's side. The patient had had hip-joint disease, and excision was performed on the right side when he was 18 years of age. He had now had sciatica and lumbago for six weeks, and was prescribed an alkaline mixture. On July 16 the pain was much the same. On the 20th the right leg was placed in the cylinder, after which the pain was much better. On the 21st the bath was repeated, with a completely satisfactory result. On the 23rd the patient had had no pain since the last bath. On August 6 he continued to be quite free from pain, and had returned to work.

CASE 8.

Neuralgia following Herpes.—The patient, who came under notice on July 13, 1896, was a woman aged 65 years. She was stated to have had influenza two years ago, which was followed by dropsy in the hands and feet, with which she was laid up for twelve weeks. Ten days previously to her coming under treatment an eruption came out over the left side of the neck, which followed three days' throbbing in this region. The case was rather a severe one of herpes; the blebs were unusually large and confluent over the left side of the neck, extending on to the face in front of the ear, up behind the ear, just into the hairy scalp, and slightly over the clavicle. She was given some starch and boracic powder to dust on, and a quinine mixture to take internally. There was albuminuria. On the 20th she was going on well, the eruption having nearly disappeared. On the 27th it was noticed that for the previous five days the patient had suffered severe neuralgic pain all over the region of the eruption; this had prevented her from sleeping at night; she was apparently in great pain and much depressed, and was crying. On the 28th the pain was just the same. The left arm up to the shoulder was placed in one of the hot-air cylinders. In forty minutes all the pain had gone, and, although some returned the same evening, she had a very fair night. On the 29th there was still some pain, but not so bad; the bath was repeated, during which process the pain again quite disappeared. On the 30th there had been little or no pain since the bath on the 29th, and the patient felt much better, and had had a very good night's rest. On August 6 there had been no return of the pain.

CASE 9.
Treated at the Liverpool Infirmary.

James W. states that about five weeks ago he felt pains gradually spreading from left hip to left knee; went on working for about a fortnight, but had to lie up at home for a week; getting no better, he came to this hospital August 4, 1896. Was treated medicinally until August 18, when the Tallerman treatment was commenced. The patient had eleven baths, having one every other day. His highest temperature during bath was 100°; highest temperature of bath, 245°. Left hospital, September 19, quite well.

JOSEPH MAGUIRE, Res. Med. Officer.

CASE 10.

Shown at the Brighton and Sussex Medico-Chirurgical Society by DR. HEDLEY.

Paralysis and Tactile Insensibility following Injury.—J. B., aged 35, states that in September, 1892, he received some very severe ill-treatment, consisting chiefly of blows on the face and head. He was taken to the Sussex County Hospital, where he remained unconscious for thirty-six hours. It appears that the entire right side of the body was found to be paralyzed and anæsthetic.

His condition much improved under treatment, and he was discharged December, 1892, attending as an out-patient up to May, 1893. Since then he had not had treatment excepting 'penny shocks from an electrical machine,' until June 26, 1894. On that day—*i.e.*, one year and nine months after the accident—he came under the observation of the writer. His condition was then as follows:

He said his hand and leg were both getting worse, and he was altogether 'going back.' His right leg was cold, and in walking perceptibly 'dragged.' Knee-jerk exaggerated on that side. 'Squeeze' of right hand, 40 lb.; of left, 55 lb. There was some contracture of flexor tendons of fore, middle, and ring fingers. He said the hand felt 'stiff' and 'numb.' He could not pick up pins. Tactile sensibility much impaired. Of the other 'sensibilities,' those of pain and temperature were both diminished. Muscular sense did not seem to be impaired. The electrical reactions presented no feature of interest. Under general electrical treatment, especially by the alternating current bath, motor-power improved, and under faradic brushing tactile sensibility began slowly to recover itself. In view, however, of recent experiences, it was considered probable that in the application of hot air at high temperatures would be found a means of powerfully aiding the action of the faradic brush in stimulating the receptive organs situated in the skin. The affected hand was therefore placed in the cylinder for forty minutes at a temperature of 160° F., which on six occasions was gradually raised to 245° F., and on one occasion to 295° F. The immediate result on each occasion was, as the patient expressed it, that the hand felt less 'numb,' more 'alive' and 'supple' than before. The points of a pair of compasses could be felt at decreasing distances. The hand has become correspondingly useful. He could write better, pick up pins, etc. Besides this quickening of cutaneous sensibility, the hot-air process, aided by manipulation, has materially relaxed the contractions, and the writer cannot help thinking that it has also in other ways improved the motor-power of the limb.

CASES 11 AND 12.*

Shown at the North-west London Clinical Society by DR. SIBLEY.

CASE 11.

This patient had suffered from malnutrition of the leg after ligature of the femoral artery and injury to the nerve. He was in the hospital under Mr. Durham from September, 1895, to July, 1896, with an abscess in the popliteal space which eroded into the popliteal artery. This was ligatured, and afterwards the femoral had to be tied on account of hæmorrhage. There appeared also to have been considerable injury to the popliteal nerve. When he first came under Dr. Sibley's treatment in October of last year, the leg was cold, very blue, and much withered. There was considerable impairment of sensation in places, especially about the foot, complete anæsthesia of the great toe and inner side of the foot; there was also perversion of sensation, the ankle was very stiff, and the patient could hardly walk. By March 15 the patient had had eighteen baths. The limb was much stronger, and he was able to get about comfortably; the feeling in the foot was much improved, and the limb remained constantly warm. The general nutrition of the part had also improved.

CASE 12.

The next case was one of malnutrition of the hand, due to severance of the nerves of the wrist. The patient received a severe cut across the wrist-joint in October last year, which was followed by loss of sensation in the last

* From the *Clinical Journal*, April 28, 1897.

three fingers, and the hand could not be used for anything. The first bath was given on January 4, and the patient stated that some feeling returned after the second bath, and remained for a few hours, but was lost again afterwards. On February 16 he had had six baths, and was able to use his hand a little. Sensation had also returned to a considerable degree in the fingers.

CASE 13.*
Treated at 50, Welbeck Street.

Writer's Cramp.—It was kindly meant, but it grew a trifle monotonous. I could not meet an acquaintance, or receive a letter, without being confronted by the sympathetic inquiry, ' How is your poor arm ?' And as the ' poor arm ' was never a ha'porth the better, and I had gradually grown philosophical on the subject, I should almost have preferred—not quite, perhaps, for human nature has a secret craving for the very sympathy it affects to despise—still, I say, I could almost have preferred others forgetting what I had myself ceased to think about. The poet says :

> ' 'Tis well that man to all the varying states
> Of good or ill his mind accommodates;
> He not alone progressive grief sustains,
> But soon submits to inexperienced pains.'

Even so had the present writer come to ' submit,' and would probably have gone on ' submitting ' to the end of time, but for the piece of good fortune now about to be narrated for the benefit of others in a like case, who have calmly surrendered themselves to what they feel to be the inevitable, but who may now take heart of grace, and look forward to a deliverance as unexpected and extraordinary as my own.

Let me, therefore, tell my tale. About three years ago, after nearly twenty years of busy authorship, and at the close of a prolonged strain, I became aware of a somewhat sudden failure of strength in my right hand and arm, joined to uneasiness in the joints of the elbow and wrist, which sometimes extended even to the shoulder. This increased as time wore on, and under medical advice a course of baths and waters, first at Bath and then at Buxton, was undertaken; but although my health was benefited, and, indeed, was never greatly affected by the local ' worry,' the writing-arm was untouched. I then

* From the *Court Circular.*

sought the advice of various well-known authorities—there is no need to mention names : I tried them all—not excepting *the* one, whose name will rise at once to your lips, and though he did indeed effect what none of the rest had, a partial cure, enabling me to release the now thoroughly weakened and enervated arm from its sling, and to use it for every other purpose than the one for which it appears to have been primarily intended—*that* remained unattainable as ever. I could not do any literary work whatever without the aid of a secretary. Furthermore, I had fits of severe neuralgia in the shoulder-blade of the *other* arm, occasioned, I was told, by sympathy. (N.B.—A curious and not altogether satisfactory mode of evincing such. I do not recommend it.)

This state of things having gone on for two years, it was time for some fresh development to take place, and accordingly, last November, during a wet week in which everyone complained of more or less rheumatism, I started a stiff knee, which was painful whenever unbent, made me walk lame, and dread going upstairs. In connection were divers lesser rheumatic aches and pains.

'Are you not doing anything to get rid of your rheumatism?' observed a friend one day. 'It appears to grow worse, and with the winter before you, it is rather a gloomy prospect. What do you say to trying the new hot-air cure?' What I said need not here be related. Fill it in, dear reader, for yourself, for I was marvellously disinclined to try any new cure, whether of earth, air, fire, or water, just as disinclined as you are, perchance, and every whit as sceptical. My friend—he is one whose opinion I greatly value—heard me out patiently, and then argued the point. Finally, I agreed to glance at the little book on the invention, and having done so, the day was his own. If 'she who hesitates is lost,' he or she who reads the varied and marvellous accounts of cases treated by Mr. Tallerman, testified to by some of our highest medical authorities, has no loophole of escape left for unbelief, or indolence, or neglect. The cure for rheumatic affections is to *be had*, and those who wish to be cured *can* be, if they choose.

This fact impressed on my mind, I presented myself one autumnal afternoon at 50, Welbeck Street, having previously made an appointment, and then and there had my first so-called 'bath.' Let me try to explain what took place. It was so very much less formidable than I had anticipated, that the narrative may encourage others as doubtful, as shrinking as myself.

I was shown into a pleasant back-room, with a skylight window and a blazing fire, where ranged round about were eight or ten large copper cylinders—somewhat resembling miniature railway engines—presided over by a bright young nurse, who at once conducted me into a small adjoining dressing-room, bade me disrobe, and rehabille myself solely in the flannel dressing-gown brought for the purpose. That done, I stepped forth again into the 'railway-shed'—how cosy and cheerful it looked, with all its shining brass and copper, that dismal afternoon!—and Mrs. Nurse at once proceeded to seat me in a comfortable arm-chair wheeled up to the mouth of one of her cylinders, already prepared for the purpose. She then carefully placed the limb to be treated within the copper chamber. The cylinder was then closed at the other end, and I was encased in blankets, given a tumbler of water to drink (either hot or cold, according to taste, equally promotes the action of the skin), my pulse was felt, my temperature taken, and the operation had begun. The thermometer, indicating the gradual increase of heat within the cylinder, rose and rose; the patient grew very warm (without, however, experiencing the slightest pain or uneasiness), and three-quarters of an hour passed.

Throughout the whole period of the bath, and even when the thermometer attached to the hot-air chamber stood at 250°, there was only a comfortable glow in the veins, with a pulse increased by a dozen beats, and a temperature sent up two degrees!

For be it observed that the *whole system* is affected through the member—whichever it be—which is encased in the heated chamber, while yet the patient, breathing a cooler atmosphere, experiences none of that lassitude and inclination to headache familiar to those who have undergone courses of Turkish and hot-air baths. It will also be remembered that a *moist* atmosphere of 115° degrees is the highest that can be endured. I repeat, then, that I experienced neither weariness nor discomfort during the hot-air application, and only a lively curiosity to know the result.

And this was the result—that the stiff knee, on being withdrawn from the cylinder, bent and unbent, and was moved and twisted about by skilful hands, without the slightest difficulty, and without a twinge of pain.

I then quickly dressed, and after remaining a short time in the outer reception-room in order to cool down before facing the chill, misty, open air, walked away from the door with greater ease than I had done since the attack began. I was not completely cured, for, though much lessened, I was conscious of a return of the difficulty in straightening the joint, and accordingly another bath speedily followed, this second application being attended with such absolute and unqualified success that I have never given the troublesome member a thought since. That took place in the middle of December last.

Encouraged now to try issues with the hardened offender —my 'poor arm'—which still continued to receive its usual quantum of sympathy and condolences, I agreed to make a trial on its behalf, although with the secret reservation that whereas it was comparatively easy to overcome a mere passing ailment in a perfectly sound and healthy limb, it would be another matter altogether to restore one to use and strength which had already resisted the efforts of some of the ablest medical men of our time.

I should, of course, be patient and just, but if at the end of a course of six or seven hot-air baths—well, it would only be the old disappointment reappearing in a new form, and, prepared for resignation, I began the course.

As the application was precisely similar to that already described, I need not repeat what now took place. The arm, in place of the leg, was encased within the heated cylinder, and that was all.

Anxious to give the experiment every chance, I arranged to take the baths on consecutive days, and after the first four, so marked was the improvement that it almost seemed as though the cure were actually effected. My fingers, which had been rigid, could close easily, wrist and elbow were supple as in days of old, and I could not only write, but write with a hand that did not ache. It took, however, three more baths—seven in all—to exorcise the fiend, as I humbly hope for ever.

Nor is this all—the treatment for my 'poor arm' has operated favourably upon my whole system. I was slightly rheumatic all over. Since undergoing the hot-air cure I have not felt a twinge of rheumatism, albeit this is the time of year at which I am most subject to aches and pains. I feel better in every way, refreshed, strengthened. I no longer employ a secretary, but am fulfilling my literary engagements with my own hand, and it seems to me that I can do no less than make known to the many who have condoled in the past my claim to congratulation in the present;

whilst in the hope that to other rheumatic sufferers this idea of deliverance, which came as a revelation, may prove a blessing, I append my name in full to this true statement of facts.

L. B. WALFORD.

CASE 14.

Shown at the North-West London Clinical Society, November 18, 1896.

Chorea Major, Rheumatism, Endo- and Peri-carditis.—Dr. Knowsley Sibley showed a boy, aged 13, who was admitted into the hospital under his care on October 15. He had been ill a fortnight, and was unable to stand, suffering with very severe choreic movements of the whole body, especially the muscles of the face and upper extremities. The head was thrown back, pupils widely dilated, lips were pouting; there was considerable respiratory embarrassment; inspiration was fairly good, but there was great difficulty in expiration, several feeble attempts preceded a violent expiratory sigh. The tongue was protruding from the mouth, and almost bitten in half. The temperature was 103°, and pulse 120. There was a faint systolic apex murmur, accompanied by a slight pericardial one. The next day the pericardial friction had considerably extended, and could be heard all over the cardiac region. There was also great difficulty in deglutition. The boy complained of some pain in the knees; there was no effusion or redness of these. By the fifth day the temperature had dropped to normal, and did not rise again, and nearly all the movements except those of the forehead had subsided. Now, at the end of a month, the child was, with the exception of the mitral regurgitation, well. The treatment had been a mixture of salicylate of soda and digitalis every six hours, and dry hot-air baths. On November 3 he was put on strychnia and iron tonic.—*Clinical Journal,* December 16, 1896.

CHAPTER VIII.

SPRAINS AND OTHER INJURIES.

It is well known that few minor injuries are so troublesome as sprains, and that they have become extremely frequent since the spread of bicycling. To those who live by manual labour they are an extremely serious matter, and, of course, they are also very common. A rapid cure is of the greatest importance, not only to the workers themselves, but to their benefit societies, and, under the new legislation, to employers of labour. Their attention is therefore drawn to the value of the Tallerman treatment in this class of cases, as evidenced by the following notes:

CASE 1.

Treated at the North-West London Hospital. Under the care of MR. MAYO COLLIER, F.R.C.S.

W. H.; aged 11. Sprain of right ankle. Right ankle and foot swollen, tender, and painful, with great pain on movement and inability to put the injured foot to the ground. There was a good deal of effusion, both into the tendon-sheaths and into the joint itself.

First Operation, August 13. *Forty minutes.*

On examination, after treatment, the pain was found to have almost disappeared; there was considerably less effusion, and the foot could be put to the ground.

Second Operation, August 14. *Thirty-four minutes.*

Further improvement, with power to flex and extend the foot.

CASE 2.—Showing ability to stand with the whole weight of the body on the injured foot.

To face p. 117.

Third and last Operation, August 16. *Thirty minutes.*

The effusion entirely absorbed; patient walked well and without pain, and was discharged as cured.

Note.—The result of the treatment in this case was very satisfactory, a severe sprain having been cured after three operations in four days.

J. F. SARGEANT, M.R.C.S., L.R.C.P.

CASES 2 AND 3.

Treated at the North-West London Hospital. Under the care of MR. MAYO COLLIER, F.R.C.S.

T. Y., aged 22, butcher; W. J., aged 24, groom. Both these cases were sprained ankles, with the usual severe symptoms. Each was treated on five occasions with most satisfactory results. The pain in both cases had disappeared after the second operation, and the joints were quite sound after the fifth.

CASE 4.

Treated by DR. ÉDOUARD CHRÉTIEN.

Contusion and Sprain of the Right Elbow.—Bernard L., aged 33, sustained by a fall from his bicycle a severe contusion and sprain of the right elbow. Flexion and extension of the forearm were impossible. The movement of the fingers caused a severe pain in the elbow-joint. Unable to write. Functional impotence.

After the second Tallerman bath the patient wrote without difficulty; after the third the movements of the fingers no longer caused pain.

CASES 5-13.

Treated at the Liverpool Infirmary.

CASE 5.

Mark M., aged 50, fell down some stone steps on June 19, 1896, and sprained right ankle. Admitted to the hospital June 21. Ankle swollen and painful; treated with evaporating lotion and rest in bed till July 11. Bath commenced. Had in all sixteen baths, his highest temperature being 99·6°, and the highest temperature of bath 235°. On July 29 a blister found compelling patient to keep to bed again. He took his discharge August 22, quite recovered. Duration of bath sixty minutes.

Case 6.

Patrick R., aged 35, fell off a parapet and sprained left ankle. Admitted to hospital July 4, the night of the accident. Treatment by Tallerman process commenced July 9. Had five consecutive baths; able to walk after first bath. Patient's highest temperature 99·8°; highest temperature of bath 240°. Left July 16, quite well.

Case 7.

James J., aged 67, states he slipped and fell on July 6, spraining his ankle. Admitted to hospital same day. Rest in bed and evaporating lotion treatment adopted till July 20, when bath commenced. Had five baths. Patient's highest temperature during baths 99·8°; highest temperature of baths 215°. Left hospital quite cured of all pain.

Case 8.

William M., aged 59, had his hand crushed between casks on April 27, 1896, taking away the first joint of thumb. He was admitted to the hospital April 30, when the entire hand was found to be very swollen and very painful. It was lanced on several occasions, and free exit given to any pus. Boracic baths were constantly employed. Improved gradually. Superheated dry-air baths commenced August 14, to relieve stiffness and reduce size of joints. This patient had fifteen baths, his highest temperature during baths being 99·8°; highest temperature of baths 265°. He left the hospital September 14, having regained the use of hand and fingers.

Case 9.

Patrick F., aged 56, states he sprained his ankle by falling on January 16, 1897. Admitted here January 19. Ankle swollen and very painful. Evaporating lotion and McIntyre splint. Tallerman bath started February 6, and finished on the 16th. Five baths were given. Patient's highest temperature in bath 99°; highest temperature of bath 260°. Left hospital February 22, 1897. Walked well; no pain.

Case 10.

Thomas W., aged 24, sprained his ankle by falling while at work. Admitted here January 22, 1897, and limb placed in McIntyre splint. Evaporating lotion applied to ankle. Baths commenced February 8 and finished March 2,

Case 11.

Robert W., aged 38, admitted suffering from sprained ankle, received January 27, 1897. Came here February 1, and commenced bath treatment February 5; had another bath February 13, and left hospital February 15, quite cured. Patient's highest temperature in bath 98·4°; highest temperature of bath 255°.

Case 12.

Gabriel J., aged 44, had ankle poisoned about four months ago; after wound had healed ankle commenced to swell and feel painful. General health not good. Had two baths—one January 27, 1897, the other on January 30. Patient felt much relief. His highest temperature in bath 101·2°; highest temperature of bath 250°.

Case 13.

Samuel S.; aged 24. Admitted February 5, 1897, suffering from a stiff and painful foot, the result of a nail penetrating the sole. Had three baths. Greatly improved. Patient's highest temperature in bath 99·4°; highest temperature of bath 270°.

JOSEPH MAGUIRE, Resident Medical Officer.

Case 14.

Shown at the Brighton and Sussex Medico - Chirurgical Society by DR. HEDLEY.

Sprained Wrist.—On September 28 a conductor in the employment of the Brighton Omnibus Company fell from the top of his bus, and injured his wrist. On the following day he presented himself for treatment. The wrist was red and very much swollen; severe pain was complained of up to the elbow, in consequence of which, during the previous night, he had no sleep. No fracture or dislocation could be detected. The hand and forearm were placed in the superheated dry-air cylinder for forty minutes, at a commencing temperature of 160° F., reaching gradually 240°. During the operation the man volunteered the statement that his wrist and fingers were getting more movable and less painful. On being released from the cylinder, he said that 'three parts of the pain had gone.' He was to return on the following day. He did not do so; but on being seen at work on October 2, was questioned, and he explained that he did not come simply because his wrist was practically well on the day of the bath, and quite well now. It would have been more prudent, however, to have had it treated again at the time named. This case presented the appearance of a really formidable sprain; its rapid cure was very striking, and the result of the treatment, both in point of quickness and completeness, seems in this instance at least far superior to that of any method with which the present writer is acquainted. Had there been fracture or dislocation as well as the symptoms named, it seems not improbable that an application of this kind would have proved a useful preliminary to reduction and fixation.

Fracture and sprain from bicycle accident; before treatment.

Fracture and sprain from bicycle accident; after treatment: showing ability to close the hand.

CHAPTER IX.

CHRONIC ULCERS.

ONLY a few cases are as yet available for publication, but they are of a very encouraging character. The case of varicose ulcers treated in Paris, and illustrated by photographs, is interesting, because it shows the vast superiority of the hot-air treatment over the ordinary boracic acid compresses. Under the former the ulcer on the front of the left leg cicatrized over without any enlargement, while the ulcer at the back of the right leg, treated by boracic acid, shows an enormous extension of morbid infiltration, coming right round to the front of the leg, although it was originally the smaller of the two.

CASE 1.

Treated at the Laennec Hospital, Paris. Under the care of PROFESSOR LANDOUZY.

P. C. Varicose ulcers : (1) on front of left leg, and (2) at back of right leg ; one year's standing. (1) Treated by Tallerman apparatus : after ten baths completely cicatrized. (2) Treated by compresses of boracic acid : also cicatrized after thirteen days, but the skin is exceedingly thin on the face of the ulcer, and all round is a mass of œdematous infiltration.

CASE 2.

Treated at the North-West London Hospital. Under the care of MR. MAYO COLLIER, F.R.C.S.

E. W. ; aged 45. Staymaker. Suffered from chronic ulcer of right leg for five years. Treated at North-West London Hospital as out-patient without improvement, and was in-patient at the Temperance Hospital for six weeks, at the end of which time the ulcer was healed. Two years ago it broke out again after a fall, and has since been getting worse.

On examination, a large unhealthy ulcer, about the size of the palm of the hand, was found in front of right leg, just above ankle; edges unhealthy, base covered with thin white slough. Loss of tissue about $\frac{3}{16}$ inch in depth.

Treated twice in superheated dry-air bath, each operation averaging thirty-five minutes. The base was then found to be much cleaner, and the margins showed signs of healing. About fourteen days later patient seen by Mr. Mayo Collier, who found that the lost tissue was replaced, the ulcer had filled up, the margins showing further unmistakable signs of healing.

J. F. SARGEANT, M.R.C.S., L.R.C.P.

CASE 3.

Shown at the North-West London Clinical Society by
DR. SIBLEY.

This case was one of chronic ulcer on the leg of a man, aged 60, who came under his care suffering from general bronchitis and dyspepsia. The ulcers, two in number, had existed more than a year, and were situated just below the left internal malleolus, being about 1½ inches in diameter. There was a good deal of general œdema of the part, and the leg was very painful. The patient was ordered some boracic ointment; and as the ulcers were getting larger instead of smaller, he was ordered the Tallerman treatment on February 2. After two baths it was obvious that the ulcers were healing, and the healing was complete on March 1, by which time the patient had had ten baths. The general nutrition of the whole limb had also considerably improved, there was less œdema, and the inflammatory area around the ulcers had disappeared. He considered that the hot-air treatment compared very favourably with the oxygen treatment for ulcers and other diseases.—*Clinical Journal*, April 28, 1897.

Case 1.—Before treatment, showing ulcer in front of left leg.

To face p. 122—I.

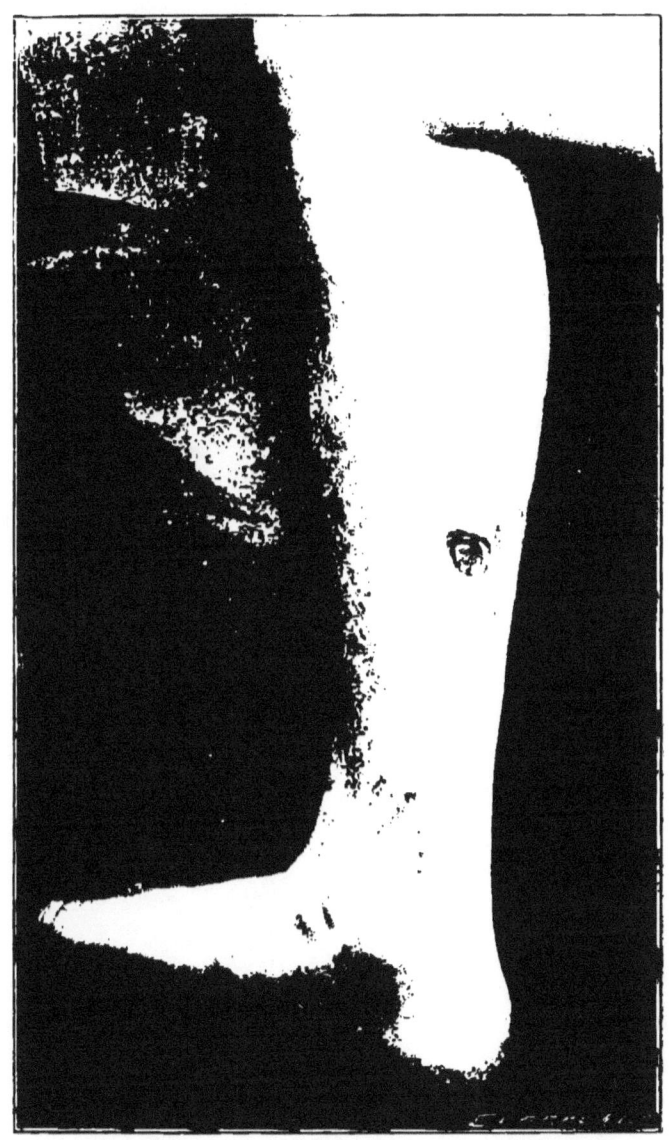

CASE 1.—Before treatment, showing ulcer at back of right leg.

CASE 1.—After treatment: Right leg treated by boracic compresses; ulcer spread round to the front (see next photograph); left leg treated by Tallerman apparatus; ulcer healed.

Case 1.—After treatment by boracic compresses: showing enlargement of the ulcer.

CHAPTER X.

TREATMENT OF RIGID DEGREES OF FLAT-FOOT BY THE TALLERMAN LOCALIZED SUPERHEATED DRY-AIR BATH.

By W. J. WALSHAM, F.R.C.S.

For the last fourteen years I have been accustomed to treat the numerous cases of rigid flat-foot that came under my care in the orthopædic department of St. Bartholomew's Hospital either by manipulation and massage or by wrenching under an anæsthetic. For the last six months or so I have substituted for this treatment the use of the Tallerman Superheated Dry-air Bath. The class of cases most suitable for manipulation and massage are those in which, on taking the foot in two hands and making pressure on the head of the astragalus with the ball of the thumb, whilst at the same time adducting and inverting the front of the foot, the muscles gradually yield, and the foot can be made to assume its normal shape, the arch being completely restored for the time being. This manipulation, however, is attended with acute pain, and on relaxing the pressure the foot at once resumes the deformed position. By repeating the manipulation, which on each application becomes less painful, daily, or, better, several times a day, for a week or two, the rigidity can generally be overcome, especially if the patient can take rest during the treatment. The foot is then in a condition for the flattening to be permanently cured or relieved by exercise, combined with some form of valgus boot. For this variety of rigid flat-foot the hot-air bath is most useful. After the rigid foot has been in the bath at a temperature varying from 270° to 280° F., or higher, for half an hour to three-quarters of an hour, it comes out quite supple, and on taking it in the hand, the arch can be restored without any pain or the application of any force. After a few hours the rigidity gradually returns, and the process, like the manipulation, has to be repeated. The bath does not, of course, cure the flat-foot, but merely, by getting rid of the rigidity, places it in a condition to be acted upon by the exercises or other means employed for its permanent relief. The advantage of the bath treatment over the manipulative consists in the entire absence of pain attending it. The bath should be employed once or twice daily, the patient resting in the intervals, until the rigidity has disappeared.

There is a more severe degree of rigid flat-foot, however, in which the foot cannot be restored by manipulation, but in which, nevertheless, the foot yields more or less completely to wrenching with the patient under an anæsthetic. No hard and fast line can, of course, be drawn between these two degrees, or, indeed, between the degree in which the foot can be wrenched into shape under an anæsthetic, and the still more severe form of rigid flat-foot in which osseous changes have occurred, and in which nothing less than an operation on the tarsus will suffice to restore the arch. This last class of cases is, in my experience, very rare. I have only met with two or three cases in the last fourteen years in which I have found that the foot could not be corrected by wrenching, and in which I have consequently been driven to more severe measures, such as Ogston's operation. In the rigid cases (third degree, or advanced flat-foot) in which manipulation does not suffice, the Localized Hot-air Bath is also useful, but only for the less severe degrees of this variety. For the cases bordering (if such an expression may be used) on osseous deformity I have found an anæsthetic still needful. After the hot-air bath has been employed in the milder forms of these rigid degrees that will not yield to manipulation, the foot does not come out supple, as it does in the cases that yield to manipulation, but on taking it in the hands it can, in many cases, be easily made to take a better position, adhesions being felt to give as when the foot is wrenched under anæsthesia. The wrenching, however, after the bath is not quite devoid of pain, but the pain is usually much less severe than that which is caused by manipulation of the variety that yields to this method. In a few very severe cases not much, if any, good has been obtained from the bath, although it has been repeated on successive orthopædic out-patient days for many weeks.

For those who are not acquainted with the bath, it may be said that it is an ingenious apparatus for keeping the air really dry, so that a temperature from 250° to 300° F. can be borne for an indefinite time. It consists of a copper chamber, varying in size and shape according to the part it is desired to treat, but generally taking the form of a cylinder. The foot, knee, or the whole leg, including the hip, may be placed in it, and similarly the hand, forearm, or whole arm. The limb, or part of the limb, to be treated is passed into the open end of the cylinder through an air-tight curtain, which is afterwards secured in such a way as to close the chamber completely. The distal end of the cylinder is furnished with an ingenious arrangement which plays an important part in keeping the air dry. The heat is supplied externally by gas-jets or oil. The temperature is indicated by a thermometer, the bulb of which is inside the cylinder at the level of the limb, whilst the scale passes outside, where it can be read off. The cylinder is furnished with a double stopcock, which can be connected with an air-pump and used either for drawing off heated vapour or admitting medicated vapour. The limb rests at the bottom of the cylinder on a metal cradle protected by asbestos, which prevents all danger of scorching the skin by contact with the heated metal. The patient can lie in bed or be seated in an arm-chair during the treatment. The most convenient way of using the bath is to heat the cylinder up to 150° F. before inserting the limb, and then gradually raise the temperature, the process of drying the air being frequently repeated, which enables the patient to bear exposure to a temperature of 250° to 300° F., or even higher, without discomfort.

This striking result seems due to the system of ventilation employed, since each time it is brought into play a considerably higher temperature can be borne without incommoding the patient, until the air again becomes charged with moisture evaporated from the skin. The plan of keeping the air dry is the distinctive feature of the bath.

It may here be mentioned that the physiological effects of the bath are: (1) Profuse perspiration of the part treated. (2) Increased flow of blood through the skin of the part. (3) Softening and relaxation of the tissues. (4) General perspiration over the whole body, and a raising of the temperature of the blood one or two degrees.

I have used the bath in many other affections than flat-foot, as stiff and painful joints, sprains, rheumatic affections, etc., but with these the present paper is not intended to deal.

I now have notes of a number of patients with varying degrees of rigid flat-foot who have been treated by the bath. A few illustrative cases will suffice:

Case 1.

Rigid Flat-foot Reducible on Manipulation.—A. B.; aged 22 years. Stands all day; both feet very rigid and painful; exercise cannot be done; has been wearing a valgus-boot for months.

February 25, 1895.—Bath 260° F. for twenty minutes; both feet came out quite supple and completely restored.

March 4.—Less pain; feet again rigid; bath 260° F. for three-quarters of an hour; feet quite restored; all rigidity gone.

March 11.—Still unable to do exercise; much less pain in feet; some rigidity; bath 250° F. for half an hour; feet perfect; suppleness returned; placed in plaster in corrected position.

March 18.—Feet very pliable in every direction; have retained their pliability; can do all valgus exercises perfectly. To continue exercise and wear simple valgus-boot.

Case 2.

Rigid Flat-foot Reducible on Manipulation.—C. D.; aged 14 years. Both feet flat, painful and rigid; cannot do exercise.

February 25.—Bath 290° F. for half an hour; left foot well restored, soon becoming rigid again; right foot only partially restored; left foot put in plaster in restored position.

March 11.—Bath 260° F. for fifteen minutes; left foot quite supple and restored; right foot better; can be almost restored on wrenching.

March 18.—Can stand on tip-toe and do exercise very well; right arch not so good as left. Bath 210° F. for half an hour.

March 25.—Does exercise very well; right foot still improving; bath.

April 8.—Both feet perfect; arch and suppleness retained; to continue exercise; baths to cease.

Case 3.

Rigid Flat-foot Irreducible on Manipulation.— G. H.; aged 32 years. Both feet very rigid and painful; incapacitated for work.

February 18.—Bath 250° F. for one hour; both feet became supple and arches restored.

February 25.—Feet soon became rigid after last bath, but both were found less rigid than before the bath. Bath 250° F. for half an hour; feet perfectly restored; put in plaster.

March 4.—Feet still slightly rigid. Bath 260° F. for half an hour; feet quite supple; exercise begun; plaster.

March 11.—Less rigid. Bath 280° F. for half an hour; feet quite supple; exercise better done.

March 25.—Feet much improved; left foot very good. Bath 250° F. for half an hour.

April 1.—Still progressing; does exercise fairly well; rigidity almost gone; fitted with valgus-boot; exercise to be continued.

Case 4.

Rigid Flat-foot Irreducible on Manipulation.—C. J. B.; aged 22 years. Right foot very rigid; left foot scaphoid and astragalus extremely prominent, probably bony changes.

February 18.—Bath 235° F. for half an hour; right foot completely restored; left foot unaltered; cannot be wrenched into position; both feet placed in plaster of Paris.

February 25.—Feet kept in plaster; very rigid; no bath.

March 11.—Right foot in good position; left same as before. Bath 250° F. for half an hour; left foot only; prominence of scaphoid slightly reduced.

March 18.—Right foot much improved; left foot better, but bones still prominent; bath 240° F. for half an hour; bones further reduced; whole foot more supple.

March 25.—Left foot more pliable, but still rigid. Bath 240° F. for half an hour; bones incompletely reduced; from this date the foot further improved, but was not completely restored.

Case 5.

Rigid Flat-foot Irreducible on Manipulation.—S. S.; aged 40 years. Rigid flat-foot of some years' duration; both feet astragalus and scaphoid very prominent; thickening about Chopart's joint.

May 20.—Bath 270° F. for half an hour; pain less; foot could not be reduced; rigidity slightly less.

May 27.—Very little improvement. Bath 270° F. for half an hour.

June 3.—No improvement. Bath 270° F. for half an hour; little or no improvement; plaster of Paris.

June 10.—Feet still rigid; wrenched under an anæsthetic; improvement in position; plaster of Paris.

Feet never completely restored, but the pain and disability much relieved; to wear valgus-boot and leg-irons.*

* From the Transactions of the American Orthopædic Association, vol. viii.

A casual at the Free Institute.

To face p. 120.

CHAPTER XI.

FIRST ANNUAL REPORT OF THE TALLERMAN FREE INSTITUTE, THE MISSION HALL, BLACKFRIARS, FOR THE TREATMENT OF THE NECESSITOUS POOR OF THAT DISTRICT, BY THE TALLERMAN LOCALIZED SUPERHEATED DRY-AIR METHOD (PATENT).

WITH the consent of the proprietary company and the kind assistance of a local committee and honorary medical officer, Mr. Lewis A. Tallerman opened a small institute at the above address on March 31, 1896, James Pascal, Esq., presiding.

The committee gladly welcomed the assistance and means to relieve and benefit the suffering humanity in the district, so kindly placed at their disposal by Mr. Tallerman, and provision was made for carrying on the work amongst the poor and destitute afflicted with gout and rheumatism, sciatica, lumbago, sprains, and other similar diseases, in the hope that many would be rescued from their sad and hopeless and crippled condition of pain and poverty, and, with health restored, become again able to perform the duties of life, and earn the daily bread for themselves and families.

I have now the pleasure to report these hopes have been in some measure realized, and when the circumstances which surround the sufferers are considered, the want of proper nourishment, warm clothing, and their not too sanitary housing, it must be considered in the highest degree satisfactory that so much benefit has been obtained in cases of such a chronic, intractable, and, by other methods, incurable character.

During the year twenty-three women have been treated and 215 baths given :

Two of the women have been treated once.
Three have been treated three times.
Two have been treated four times.
Three have been treated five times.
One has been treated six times.
One has been treated seven times.
Two have been treated nine times.
Two have been treated ten times.
One has been treated twenty times.
One has been treated twenty-two times.
One has been treated twenty-three times.
One has been treated thirty times.

Thirty-one men have been treated and 139 baths given:

Eleven of the men have been treated once.
Six have been treated twice.
Four have been treated three times.
One has been treated five times.
Two have been treated six times.
Two have been treated eight times.
One has been treated ten times.
One has been treated twelve times.
One has been treated twenty-three times.
One has been treated twenty-five times.

Of the women who have been treated, I might refer to:

Mrs. P., who for several years I have visited and seen her crying with pain, with nothing to remove it. She came and was treated ten times for pain in her knees, and has been cured and free from pain ever since. She told me after her first bath she had knelt down to say her prayers for the first time for five years.

Miss L., a dressmaker, who lost her business through having rheumatoid arthritis, and was scarcely able to walk or lift a cup from the table for two years, and getting worse every day. She came and was treated thirty times, and is now able to carry a jug of water, sweep the room, and walk with a firm step.*

* Four months later the patient wrote: 'I am now able to cut my own food, to use my needle, and to wear leather boots, which I was not able to do for nearly two years before having the baths.'

Miss L.—Before treatment: showing inflammation of wrists and fingers.

To face p. 130—I.

Miss L.—Before treatment: showing limited flexion of hands and deformity of elbows.

Miss L.—Before treatment: showing swollen and fixed condition of feet.

Miss L.—After first operation : showing improvement in hands and feet.

Miss L.—After treatment: showing hands and feet restored to nearly normal condition.

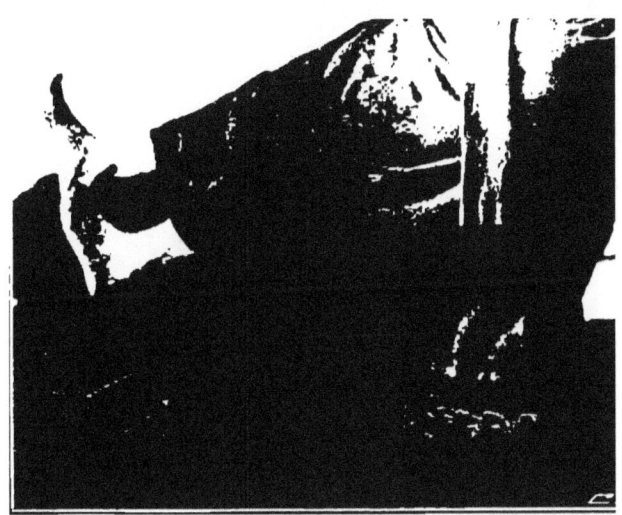

Miss L.—After treatment: showing restored movement of hands reduced swelling of feet.

Mrs. L., a tailoress, has suffered with gout in all her joints for four years after having been treated by doctors at the dispensary without any benefit; her legs, knees, finger-joints and wrists most hopelessly ruined and useless. She was treated thirty times, and is very much benefited—far more than could have been expected.

Mrs. W., my sister from Ipswich, after suffering for about two months with rheumatism in hands, legs and feet, which were sometimes very much swollen and painful, came and was treated six times in about eight days. She went home cured, and has had little or no pain since, and is quite able to do her work. She was recommended by her family doctor. She gave five shillings as a thank-offering.

Emily P., a tailoress, also came from Ipswich at the earnest wish of the Crown Street Christian Endeavour Society. She came in a bath-chair, not being able to stand. She has suffered for ten years, and has been unable to use her crutches for over four years. I permitted her to come, but held out no hope even of relief. After twenty baths it was found that all signs of sciatica had left her, and also very much pain in her limbs. Her friends were so much pleased with the results obtained that, after assisting in the cost of travelling, board and lodging, they gave eleven shillings as a thank-offering to the institution.

Frances S., a young girl aged 16, came to be treated after having had rheumatic fever, which left her with pains in nearly all her joints after having suffered eight weeks, when she applied to Dr. Hill. He wisely refused to sign her form until her heart was stronger. After two weeks she came and had five baths, and was perfectly cured, and returned to her work at Mr. Pascal's.

Alice R., another young girl, who was a servant at the Rev. Tolfree Parr's, had suffered from rheumatism for about three months, through catching cold. She came and had seven baths, and was cured.

Elizabeth M., another girl, had suffered with rheumatic fever and had been in the infirmary; she had eight baths and was cured.

Amongst the male patients we may refer to:

Mr. W., who was one of our first patients. He had been suffering from rheumatism in left foot and ankle, as well as other parts of his body, for five years. He was cured with eight baths, and has had no return since.

Mr. L., whom I found, in my visitation, lying on his back in great pain. I knew him to be subject to rheumatism for years; he had been to the infirmary, leaving his wife and children in the deepest poverty. I advised him to come and be treated. He was treated only once, and went to work the next morning; he has had a slight attack since, but has been able to work, with the exception of a few days, ever since.

Mr. S., an assistant at a fried-fish shop, came suffering with sciatica in right leg and ankle; for about three months he had been in the infirmary, and was going back because unable to work. He had two baths, and has been able to work ever since. I met him a few weeks ago, when he thanked me for my kindness to him.

Mr. W., a scaffolder, had suffered from gout in right foot and leg and great toe; he had only done three weeks' work in three months. They were very much swollen and very painful. He had only one bath, and went to work the next morning. In about a month it came on in his left arm. I gave him another bath, and he has been able to work ever since last August.

Mr. D., a navvy, came from Lewisham. He was suffering from sciatica, and unable to work. He told me the pain made him sick; he had suffered about four months, and had tried many so-called remedies, also his club doctor, with little or no benefit. He was treated ten times, and was perfectly cured. He gave five shillings as a thank-offering.

Mr. P., a brother City Missionary, had suffered with rheumatoid arthritis in his left arm and shoulder for about four months; he was unable to dress himself or raise his arm. He had been under his private doctor's treatment with no relief. I had to assist him to undress, but after his first bath he was able to dress himself, and has done so ever since. He continued to improve from the first under the treatment. He had twenty-three baths and is now perfectly cured, with the exception of a slight stiffness in his shoulder-bone, which might have been removed if he had permitted it to be wrenched, which his doctor advised him should be done; he has deferred this for the present, hoping it will come right in time. I sent him to Dr. Sibley at the North-West London Hospital, hoping he would do it for him, and he kindly treated him as an out-patient.

John H. is a box-maker at Pink's jam-factory; he was taken with gout when he was going home from his work one Saturday. He had no sleep, neither could he eat because of the pain. I saw him on the Monday, and advised him to have some baths; he replied: 'I can't get there, for I can

Rheumatism: before treatment.

Rheumatism: after treatment.

To face p 132—II.

scarcely bear to touch the floor with my foot.' I suggested that he should get a costermonger's barrow and have someone wheel him to the doctor's to have a form signed, and then come on to the mission-hall in the morning. I then said: 'I will promise you that you will be able to walk home.' He did so, and after the first bath he lost all pain and part of the swelling, and he walked home. He had another bath on the Thursday, when he lost the soreness, and went to work on the following Saturday, and has been at work ever since. This man had suffered about four years before, when he drank very heavily, and lost a great deal of time through not having anything to ease him. He has been a total abstainer for some months.

Mr. S., a corn-porter and carman, had been unable to work for twelve weeks through rheumatism in his knees. After two baths he went to work again, but needed more treatment. The pain came on again about a month afterwards, when he came and had four more baths, and has gone back to his work.

Mr. M., a carpenter, has suffered from rheumatism over six years. At the time of the strike he went to live at Aldershot for five years, and only worked about two years. He returned to London about fifteen months ago, and has only worked about eight months out of that time. He went to work after his first bath, but has had two baths since and is intending to have some more. When he came, his right arm was stiff, but now he has almost perfect freedom; he is of course most thankful for the benefit he has received. He has a wife and three children, and is a total abstainer.

There are other cases, both of men and women, who have not received much benefit, but as a rule there is a cause for it: some through poverty and having been unable to work have to remain in the same surroundings in which they became sufferers, living in damp and low rooms.

I have in every case observed the conditions upon which the use of the treatment was granted, namely, that the Institution should be free to all the necessitous poor, and that it should be quite unsectarian; no one who has applied has been refused, but all have been heartily welcomed. Catholics, Protestants, and persons who make no profession of religion, have shared the benefit without any question as to their belief or unbelief; still, as a London City Missionary and as a servant of God, I have had many blessed opportunities of religious conversation with the patients as well as with other persons I meet with, which I trust, with God's blessing, will result in their spiritual as well as temporal good. Most of the patients have been those who, after long suffering, pain, and inability to work, are in the depths of poverty, some even without bread to eat.

A meeting of the Committee of the Tallerman Treatment Free Institute for the use of the necessitous poor was held on Friday, May 7, and a report of the work for the past year was read. The Committee expressed their surprise and delight at the number of cases treated, and that so many were cured, especially so as the sufferers belong to a class who stood so much in need of help. The following resolution was passed, viz.: That the sincere thanks of the Committee be conveyed to Mr. Lewis A. Tallerman for his great kindness and assistance to the suffering poor.

Some patients at the Free Institute, with Mr. License and nurses.

To face p. 134.

CHAPTER XII.

MEDICAL REPORTS, ETC.

This chapter contains a series of public reports and demonstrations by medical men.

A number of private letters have also been received from physicians and surgeons of the highest standing expressing warm appreciation of the Tallerman treatment, but it is thought that the publication of such letters might be misunderstood, even when express permission has been given. Only public utterances have therefore been reproduced. Descriptions of the apparatus have been omitted from them as superfluous, and also the cases by which they were illustrated, these having already appeared under their proper headings in preceding chapters.

I.

A CLINICAL LECTURE

On the Therapeutic Action and Uses of the Localized Application of Dry Air heated to High Temperatures in Certain Classes of Surgical Affections.

Delivered at St. Bartholomew's Hospital on Wednesday, May 23, 1894, by
ALFRED WILLETT, F.R.C.S.,
Surgeon to the Hospital.

Gentlemen,—My lecture to-day is on ' The Therapeutic Action and Uses of the Localized Application of Dry Air heated to High Temperatures in certain Classes of Surgical Affections.'

*   *   *   *   *

It was only in January that I first heard of this invention, and I was startled at what I heard it could do. My information, it is true, was not at first hand, and probably, as usual, the story lost nothing in being repeated. But this is what a medical friend told me.

He said a complete revolution, he had heard, had been made in the treatment of all sorts of contractions and of anchylosed joints. By the action of air heated to 250° or 300° F., any such affection, even a congenital club-foot—and that was particularized—would, in the course of an hour or so, become so relaxed and pliant, that the surgeon could at once quite painlessly move it in any direction, and effect an immediate cure of the deformity by restoring the foot to its correct position.

Such is the story that was told me. I readily accepted an invitation to witness those vaunted powers put to the test; and, accordingly, on January 23, at the courteous invitation of Mr. Lewis A. Tallerman, I went to 1, Chiswell Street with Mr. Walsham and Mr. Paterson, my house-surgeon. There were some three or four other medical men present. Of the few patients collected two were selected for trial. The first was a middle-aged man with subacute synovitis of the knee. Increased heat, slight effusion, and some pain existed. This was increased on attempting to move the knee, which was held semi-flexed, having a range of active movement of only 10-15°. He could only walk, or rather, I should say, hobble, with the aid of two sticks, on the toes of the affected limb. In this condition he was placed in a cylinder like the one we have here. When taken out of the bath, after some thirty or forty minutes, the knee was straight, all pain had left him, the foot was on the ground, and he walked almost briskly out of the office. I heard that in a few days' time he returned to work. This result naturally made a considerable impression on me, for I know of no treatment that could have brought about so rapid a cure. The case, no doubt, was not only curable, but well on the road to recovery. At the time I said that in hospital I should have anæsthetized the patient, manipulated his knee, and then brought it straight ; but even so I feel sure that many days would have elapsed before he would have been well, whilst here the man had been apparently almost cured in little more than half an hour, not only quite painlessly, but by a process that one might almost call that of luxurious ease.

The second case was equally satisfactory—that of a woman of about 45, with both hands crippled by chronic gout. The fingers were all kept slightly flexed. Before going into the bath she could neither straighten nor flex them. It appeared that slowly, for upwards of a year, she had been drifting into this condition, and now she could do but little for herself. Only one hand, the right—said to be the worst—was put into the bath. After fifteen minutes she volunteered the statement that her fingers seemed to loosen, and soon after she could oppose the thumb to all the fingers, which for many months she had not been able to do. After half an hour the hand was removed from the bath. I saw her open and close her hand readily, while the left, which had not been treated, remained in the same fixed position. Of the subsequent state of this patient I have heard nothing. Both these patients were entirely comfortable all through the process. If they felt the bath too hot, air was readily, for a second or two, admitted, and then a higher temperature could always be borne. The patients perspired freely, and when the limbs were taken out of the bath they were very red and moist.

Subsequently, I was asked if I would test the efficacy of the bath at this hospital, and having obtained the consent of the authorities, I have done so, and desire now to put the results before you.

*   *   *   *   *

In offering some comments on this invention, I must say at once that I can only express a very guarded opinion at present, mainly because my experience is still very small ; yet, small as it is, I have no hesitation in saying it is distinctly encouraging. In the cases that have been treated in President and Pitcairn Wards I was anxious to try the bath in an assortment of diseases, and also to test its action purely and simply. In the next place, the process occupies roughly an hour for each patient, and some of the patients were undergoing the bath treatment for the entire two months. Mr. Tallerman kindly gave his valuable services in supervising its administration; hence, during the two available hours of the afternoon, with two baths going, it was not possible, as a rule, to treat more than about four patients a day, and of these some were under Mr. Walsham's care.

It seems to me that the points for consideration are these:

1. What is the effect of exposing a part, such as a hand or foot, to such temperatures as 250° to 300° F. (*a*) upon normal parts; (*b*) upon certain diseased tissues?

2. Is any therapeutic effect produced by such temperatures alone; and, if so, in what classes of diseases? and, again, is the result likely to prove temporary or permanent, partially or completely curative?

3. Are there cases in which air baths of high temperature can be advantageously employed in association with other methods of treatment, for example, where articular adhesions have been broken down?

4. To what depth from the surface is the direct influence of this dry hot-air bath exerted?

5. Is the body generally acted on or affected, either favourably or prejudicially, by the topical use of heated air as in this bath?

I will now, in turn, direct your attention to each of these several points.

As to the first of these questions, I have come to the conclusion that in principle this apparatus for applying dry hot air locally has on the member or limited part of the body so treated an action similar to that of the ordinary Turkish bath on the whole body, that is to say, it induces sweating—diaphoresis. I speak now only of the principle involved in the action of this heated cylinder, and not of the degree, nor of other differences, such as that in the cylinder treatment ordinary air is inspired, while in the Turkish bath the air breathed is greatly heated. In a Turkish bath the temperature in the first room—called the tepidarium—is from 113° to 117° F.; in the second, or calidarium, it is from 132° to 140° F. Although there are recorded instances of the higher temperatures being respired, still, probably the temperature attained in this cylinder— from 250° to 300° F.—is much too great to render its inspiration anything but hazardous—at all events, for such average individuals as frequent Turkish baths. But just by so much as the temperature of the cylinder is greater than that of the Turkish bath, by so much will its sudorific effects be increased. In the next place we note that the part, foot or hand, issues from the bath very much the colour of a boiled lobster; the flow of blood in the skin is obviously, and that in the subcutaneous tissues probably, increased greatly. The third effect noted is what we may term its anodyne influence. In most cases, not merely is pain relieved, but often it will be entirely removed. This is shown in many ways. The patients invariably experience relief. Then they will use the limbs with much greater freedom. Movements that excited pain before can be performed after the limb is placed in the bath without pain. Again, in cases of breaking down of adhesions under gas the patient is in great pain afterwards; but let the limb be placed in the heated cylinder and the pain is greatly lessened, as also is the secondary stiffness, due to extravasation and inflammation round the torn tissues. On the other hand, there is none of the excitement, amounting in some individuals to distress, from breathing the hot air of the Turkish bath before free perspiration breaks out. The patient with arm or leg in this cylinder seems throughout in the most absolute comfort, only now and again complaining of the heat being almost too great—a complaint which is instantly removed by manipulating the apparatus.

If such are the results observed from the baths in healthy persons, its therapeutic influence in certain affections is obvious, and, I think, clear to demonstration.

The next consideration, therefore, is, To what class of disease is this plan of treatment likely to be serviceable? I can only speak of its use in surgical affections. Of these, sprains, stiff joints—those where movement is more or less limited—callous limbs, flat-foot, gonorrhœal rheumatism, and possibly some skin affections, are the most likely to be benefited. The cases I have recorded show a fair selection of these.

On the third point, as to its being an adjunct to surgical treatment, I have only to repeat what I have already said, as to the great assistance this hot bath affords after breaking down articular adhesions, both in relieving pain and lessening the tendency to recurring stiffening. I think that, in addition to these, the bath may be occasionally useful as an adjunct to the electrical treatment of certain paralytic cases, and generally useful, I think, before massage to wasted muscles.

As to the extent of action, the cases recorded seem to show conclusively that the direct influence of air heated to this high degree does not extend much, if at all, beyond the skin primarily. That secondary effects are shown, I readily admit, such as relaxation of muscles which have been irritated or excited to tonic contraction; but for overcoming firm fibrous articular adhesions I am sure this hot-air bath gives the surgeon no direct help, or, putting it in other words, whilst increased power in active movement is nearly always gained, there is no immediate marked increase in the range or extent of motion.

On the last point mentioned, I think there is evidence that the effect of the bath is not confined solely to the part acted on, for the temperature of the patient is raised usually nearly one degree; true, this elevation of temperature alone would not prove anything, for excitement will often send up temperatures; but the entire skin becomes relaxed, and perspiration occurs freely over the body. Besides, patients unite in saying that with subsidence of pain in the part treated pain is lulled in other joints. I hope some physician will be induced to try the hot-air system in selected medical cases. By analogy I think that in rheumatic arthritis, gouty attacks in arms or legs, sciatica, lumbago, and perhaps in some spinal cases, good results might be anticipated.

In conclusion, I would summarize my remarks by saying that when employed for contractions in recent affections or subacute inflammatory diseases, such, for example, as may follow upon simple synovitis, cases, that is, which would yield readily, and without force, under an anæsthetic, I feel confident that the therapeutic action of this dry hot-air bath to the part will be both marked and rapidly curative. In permanent contractions or fibrous anchyloses the result of old-standing arthritic diseases, the direct therapeutic action is soon exhausted. It will be well that the cause and extent of the disease, so far as actual pathological, *i.e.*, gross alterations of structure, are concerned, should be ascertained at once, because if articular or capsular adhesions exist, valuable time would be wasted in attempting to cure such cases in this hot-air cylinder. Sooner or later the adhesions will have to be forcibly broken down under an anæsthetic; but such having been done, I think recovery will in many cases be hastened by the subsequent use of the heated cylinder; whilst in such permanent deformities as congenital club-foot, contracted scars after burns, and bony anchylosis, it is absolutely useless to suppose any effect whatever could result. It would be sheer quackery to advise its use in any such condition.

* * * * *

In so many surgical cases should this hot-air bath treatment prove serviceable, that I hope it will not be long ere one is available at any time in this hospital, and also that we have a nurse trained to its use.

From the 'Clinical Journal,' May 31, 1894, by permission.

II.

The Treatment of Rheumatic Affections by the Tallerman Superheated Dry-Air Apparatus

(*From the Medical Clinic of the Royal Victoria Hospital*),

BY

JAMES STEWART, M.D.,

Professor of Medicine and Clinic Medicine, McGill University; Physician to the Royal Victoria Hospital, Montreal;

AND

W. G. REILLY, M.D.,

Senior Resident Physician, Royal Victoria Hospital.

In December last Mr. Lewis A. Tallerman, of London, gave two demonstrations of the method of using his hot-air apparatus at the Royal Victoria Hospital before a large number of the practitioners of Montreal. Since then the apparatus has been in constant use in the treatment of various forms of subacute and chronic rheumatic affections. The results on the whole have been very satisfactory. Relief to pain has usually followed, and in nearly all cases there is soon noticed not only an improvement in the local conditions, but also a marked change for the better in general nutrition. In the present preliminary communication an account is given of three cases treated to a conclusion. In all it will be noticed that the results obtained are much more marked and satisfactory than by any other method at present known.

The Tallerman apparatus is made in various forms and sizes suitable for the different joints or limbs to be treated, viz., for the hand, arm, elbow, foot, leg, and knees, and there is also one for the pelvic region. The hospital model, the one mostly supplied to institutions, is constructed so as to enable the treatment of the extremities and to obtain a general effect upon the body.

The highest temperatures at which patients have been treated according to hospital reports are from 300° to 315° F., but the great value of the Tallerman treatment is the ability to administer these high temperatures over a period of upwards of an hour if necessary. Experimentally a patient had been treated for more than two hours at a temperature averaging 260° F., without any other discomfort than lassitude on the following day; the skin was not unfavourably affected or unduly sensitive, and it could be rubbed briskly with a towel when released. The therapeutic effects produced are relaxation of the part, copious and free perspiration over the whole of the body, enormously increased circulation, and raising of the body temperature from $1\frac{1}{2}°$ to $4°$ F. This last effect, so contrary to the belief hitherto held that the body temperature could not be raised by a local application of heat, is remarkable, and, it is the belief of Mr. Tallerman, will before long be shown to have a very material and beneficial effect in the treatment of diseases other than in the classes of cases which until now have been subjected to it; these are such cases as rheumatism, acute, subacute, and chronic, acute and chronic gout, rheumatoid arthritis, sprain, stiff and painful joints, gout, rheumatic sciatica, lumbago, peripheral neuritis, gouty neuritis, etc. Also before and after breaking down of adhesions and kindred complaints.

The Tallerman treatment has been demonstrated at some of the principal London and other hospitals, and during the course of the three years' supervision to which it has been subjected, it has been proved that the treatment can be safely administered with benefit even where great debility, weak action or valvular disease of the heart or kidney disease are present. Rheumatic and other pains are relieved, if not entirely removed, shortly after the commencement of the first operation; and the treatment itself is not only absolutely painless, but so soothing as to frequently lead to patients falling asleep if permitted whilst under it; hence the sleeplessness caused by rheumatic pain is relieved, and patients are able to rest at night.

BRITISH MEDICAL ASSOCIATION.

A paper on Rheumatoid Arthritis, read by Professor Stewart at the annual meeting of the British Medical Association at Montreal, September 1, 1897, concluded as follows:

'In my opinion the most valuable of all methods of treatment is the use of baths of superheated dry air, after the Tallerman method. It has been used in twenty cases of rheumatoid arthritis in the Royal Victoria Hospital during the past nine months with gratifying results.

'The apparatus consists of a copper cylinder, of various shapes and sizes. The model usually employed is sufficiently long to admit a lower limb to some inches above the knee. By means of valve taps the moisture from the limb is expelled, so that the air in the chamber is kept dry. The temperature in the chamber is kept usually from 240° to 300°. The first marked effect is copious perspiration all over the body. The pulse is increased from fifteen to thirty beats, and the temperature is usually elevated from 1° to 2°.

'In all we have treated twenty cases with the hot-air bath. In fourteen of the twenty cases pain in the affected joints was present, and of a severe character. In the great majority of the cases the relief was marked even after the first bath, and after several baths the patient, except on movement, was practically free from pain. As a result of this relief, sleep, which usually before was greatly disturbed, was possible. In addition, there was some apparent change for the better in nutrition. In spite of losing daily more than a pound in weight from the loss of fluid by perspiration, the patient usually steadily gains in weight. Gains of from three to four pounds weekly have been common. As regards the effect on the affected joint, it is various, depending on the amount of effusion and the degree of anchylosis.

'Generally a considerable increase in the mobility follows after the use of a few baths. It cannot be expected that restoration can take place in advanced cases, but before much actual destruction takes place, there is every reason to look for a decided check to the progressive character of the disease.

III.

Report by Professor Landouzy of the Laennec Hospital, Paris.

August 3, 1897.

From what I have seen of the local application of hot air by the Tallerman method and apparatus, both in my wards at the Laennec Hospital and in my private practice, I am enabled to say :

1. That the Tallerman method constitutes a real advance in the local applications of hot air.

2. That this treatment is indicated in a large number of acute, and a still larger number of subacute and chronic affections, both of the limbs and the adjoining parts. In these it seems to produce better results than the remedies hitherto in use. In particular, cases of gonorrhœal rheumatism, of joint affection in chronic rheumatism, of acute gout and sciatica, have been better and more rapidly relieved, better and more rapidly reduced, than by the old external remedies, whether used alone or in combination with drugs.

3. That the indications and contra-indications of its use require more precise definition, in order to determine the important place which the treatment deserves to take in the therapeutics of those painful affections of the joints which are so common in the course of infectious and constitutional diseases.

IV.

Report by Dr. Dejérine of the Salpêtrière Hospital, Paris.

August 7, 1897.

Among the cases of chronic rheumatism in my wards at the Salpêtrière, in which I have been able to appreciate the good effects of the high-temperature treatment, I have observed the following, in which this method has given the most remarkable results.

The patient was a young woman, 25 years of age, suffering from unilateral infectious rheumatism of the left side, the commencement of which dated from October, 1896. I saw her for the first time four months after the beginning of the illness, and it was the most severe case that I have ever seen. The elbow, wrist, knee, and ankle-joints were all greatly swollen, the sheaths of the tendons and serous bursæ much enlarged and excessively painful. She had been unable to leave her bed for four months and a half. On coming to Paris at the beginning of April she was placed under daily treatment by the cylinder at 212°, 230°, and 240°. At the end of a few sittings the pains had almost disappeared, and the swelling began to subside. At the present time, August 7, 1897, after 62 sittings, the patient's condition is greatly improved, and the joints have returned to their normal size.

I consider, therefore, that in this case the results of the treatment with superheated dry air by the Tallerman method are most remarkable, and superior to everything else that I have seen used in similar cases.

V.

Paper by Dr. Edouard Chrétien, of the Laennec and Salpêtrière Hospitals, Paris.

In December, 1895, Mr. Lewis A. Tallerman demonstrated in the clinic of M. Oulmont, with which I had the honour to be connected at that time, his method of treatment and the effect produced by his apparatus.

This apparatus, designed more particularly for the treatment of diseases of the joints, and employed in several hospitals in England, has given the best results. Its principle is to submit the affected part to a temperature higher than has up till now been attainable.

Everyone knows that heat exercises a favourable influence on pain. Hot compresses, poultices, sand-baths, vapour-baths, etc., produce their effect beyond a doubt by the high temperature attained, and everyone knows the wonderful results obtained by their use in the treatment of neuralgia, muscular rheumatism, and other forms of joint diseases.

But all these methods have their weak points. In some the temperature obtained and the duration of the application are insufficient, as in the employment of poultices and hot napkins. In others the method of application itself is defective, as in the case of Turkish and Roman baths, when the patient's whole body is exposed to the action of the hot vapour, and he is forced to breathe it for a period more or less long. The various apparati designed to obviate this last drawback—such as the box in which the body of the patient is enclosed with the head outside—have not given the results hoped for.

On the other hand, most of these methods employed for the application of moist heat are defective in principle, because one cannot pass a certain temperature, which is really not very high. The human organism can stand a much higher temperature when the heat is dry than when it is moist; and we see, for example, glass-workers expose themselves to the hot air of the kilns, which they could not stand if it were saturated with water-vapour.

With the Tallerman apparatus, on the contrary, thanks to its peculiar construction, we can apply hot dry air to large portions of the body—a whole limb, for example—and can attain temperatures in the neighbourhood of 300° F., or 141° C., or even higher. Observation has shown that in cases where several joints are affected, the hot-air bath acts not only on the joint enclosed in the apparatus, its influence is equally felt by the other affected joints, even those on the other side of the body. The effect obtained is only less marked. So when, for one reason or another, it is impossible to treat the affected part directly, we have tried, and successfully, too, to treat it indirectly, by placing the corresponding sound limb in the apparatus. The effect is not so marked, I repeat, but it makes a decided impression, nevertheless.

The duration of each bath varies from about half an hour to an hour. The temperature attained also varies greatly at each treatment. It is well at the beginning of the treatment to test the sensitiveness of the patient by letting the temperature remain at about 100° to 115° C. After that and at succeeding baths temperatures of 120°, 130°, 140°, and even 150° C., may be used without inconvenience.

When the bath is over the limb is taken out of the apparatus, dried, and wrapped up. The patient, who during the whole treatment has perspired abundantly, remains for a certain time wrapped up and resting. The number and frequency of the baths vary according to the case, its gravity, the inaction produced on the patient, and the results obtained.

When Mr. Tallerman placed his apparatus in our hands and asked us to test it, he declared that he had employed it in England to treat a great variety of diseases, such as rheumatic arthritis, gout, sprains, acute joint affections, stiff joints resulting from injuries or inflammations, gonorrhœal rheumatism, flat-foot, chronic and deformatory rheumatism. Since then the list of diseases which it will benefit has been greatly added to, and we can see still more which it may benefit.

Among them are muscular rheumatism, acute synovitis, sciatica, lumbago, tuberculosis, arthritis, certain forms of neuralgia, peripheral neuritis, etc.

We must say, however, that the apparatus we are describing can only be used, on account of its size and shape, to treat the limbs, or, more properly, parts of limbs, for it is impossible to introduce the entire leg into it up to the hip. To enable the treatment of the hip, abdomen, and spine, the pelvic apparatus has been constructed and so designed that the treatment can be applied directly to these parts. We will in a future issue describe this second apparatus, which is still being experimented with in England, and no model of which has yet reached Paris.

The interest of M. Oulmont was excited by the results obtained by the use of the apparatus in the English hospitals in the hands of men like Willett, Herbert Page, Ward Cousins, Walsham, Knowsley Sibley, and he asked me to secure for him some patients upon whom we could make a trial of the superheated air-baths.

My chiefs (*maîtres*), Professors Landouzy and Dejérine, to whom I had spoken of the apparatus, were equally anxious to authorize its employment in their clinics in the hospitals of Laennec and Salpêtrière.

I will now describe briefly the results obtained in Paris with a large variety of patients, both in the hospitals and in private practice, and will follow that with some observations and reflections on this treatment, what it has done, and the part it is destined to play.

Thus one can judge by these few cases described that the effects obtained by the Tallerman apparatus in chronic deformatory rheumatism, gonorrhœal rheumatism, sprain, sciatica, gout, and certain forms of joint disease of a more or less well-known origin, have been remarkable enough to merit publication. I speak only of these diseases because these are the only ones which we have treated in Paris, but the list of diseases treated in England is, as we have seen, much longer.

The study of the cases treated has shown that altogether the treatment has produced the following phenomena :

1. A perspiration, more or less profuse, not only on the part treated, but over the whole surface of the body.

2. A marked reddening of the skin on the part treated, indicating an intense dilatation of the bloodvessels.

3. The diminution and rapid disappearance, sometimes almost immediate, of pain, as in cases Nos. 4 and 6.

4. The restoration of movement where the functional impotence was due only to pain (cases Nos. 3, 4, 5, and 6).

5. A more or less marked acceleration of the pulse, caused evidently by the peripheral dilatation of the blood-

vessels produced, which facilitates and strengthens the action of the heart.

6. A temporary elevation of the body temperature.

How do these hot-air baths act? That is a complex question, to which it appears difficult to reply at present.

It is easy to say that the prolonged application of a high temperature produces an intense reaction in the affected part. But that explains nothing, for is there anything more obscure than what we describe by the word ' reaction '? It would only be an attempt to content ourselves with a form of word which explains nothing, as has already been done in so many other cases.

The most evident result from this treatment, both to patient and physician, is the disappearance of pain. But we have applied cold also to the treatment of pain (siphonage, ice, chloride of ethyl and of methyl, ether sprays, etc.), and warmth also (hot compresses, poultices, hot air, moist and dry, application of pointes de feu, cupping, blisters, injections of sterilized water), all acting in different ways, aim at the same result—the amelioration of pain.

If we compare all these processes, they have this in common—that, all being violent methods, they profoundly disturb the statics and dynamics of the tissues to which they are applied. I believe myself that that which we speak of so obscurely as 'reaction' in treatment for the relief of pain is caused by molecular changes in the great trunk nerves and their terminations. The excessive functional activity which is indicated by the dilatation of bloodvessels provoked by some of these processes, such as pointes de feu, spraying and hot air, seems to indicate action in the manner that I have mentioned, my object here being more to state facts than to explain them.

What can we hope for from the superheated dry-air treatment? What are the favourable and unfavourable symptoms produced by it? There are no unfavourable symptoms that I know of resulting from the application of the treatment. Thus, as was proved by case No. 7, the presence of a heart trouble did not forbid the use of the

Tallerman treatment; quite the contrary. In fact, the diminution of the arterial tension caused by the peripheral dilatation of the bloodvessels facilitated the heart action, as was indicated by the immediate acceleration of the pulse.

As far as affections of the respiratory organs are concerned, I have no personal experience. All that I can say is, that in several of the cases published in the English medical journals, the patients who presented themselves for treatment had, in addition to their joint affections, chronic bronchitis. That was so much helped by a course of these superheated dry-air baths that the treatment has been deliberately applied in cases of affections of the respiratory organs, and with success.

There remain still the effects on the renal organs. The English physicians are mute on this subject. I have carefully examined the urine of all the patients who were treated in the Laennec Hospital and found no change either in quantity or quality (sugar, albumen). That is a point which has not, up till now, been cleared up.

A priori, it would not seem that the hot-air baths could have any evil influence on the kidneys in their normal condition. It would be a benefit, on the contrary, in certain nephritic cases if they would, by diminishing the muscular tension, diminish the diuresis.

The analysis of urine has not only been directed to the discovery of modifications which could have existed before the treatment: we have also sought to discover if the hot-air baths influenced the quantity of matter secreted by the urinary organs during the twenty-four hours.

In case No. 5 the urine was analyzed after each bath. It showed only a very slight improvement in the elimination of salts, particularly chlorides. The daily co-efficient of the urine changed from 20·97 grains to 25·50 grains.

From this point of view, case No. 12 is most interesting, for here, in a case of long-standing gout, the daily quantity of uric acid eliminated rose from 57 centigrammes after the fourth bath to 89 centigrammes after the ninth. The patient being still under treatment, we cannot say now how far this elimination of uric acid will go, and when it will return to normal.

The field of diseases suitable for treatment by these superheated dry-air baths is, as one can see, very large. It comprises:

1. All the painful affections of the limbs, the diseases which attack the joints, the diseases of the muscles, the trunk nerves (sciatica, neuritis), and perhaps we can use it in the treatment of wasting diseases, diseases of the bone and of syphilitic origin.

2. Acute and chronic arthritic diseases, whether acquired (gout, rheumatic fever, gonorrhœal, tuberculous) or hereditary. I do not wish to say that this treatment can cure the deformities of old sufferers from rheumatism, their fibrous adhesions or their muscular atrophies. But I believe that it will abate them, and after treatment with the hot-air baths we can with much greater ease and with better results break the fibrous adhesions and restore movement to the stiff joints. I believe that by associating the hot-air treatment, electricity, and massage, we can, if not cure, at least considerably ameliorate the lot of these sufferers whom we see in the chronic disease wards, helpless and deformed from rheumatism for ten, fifteen and twenty years, and restore again the charm to their life, something which no therapeutic agent employed up till now has been able to accomplish.

As far as the joint affections are concerned, and more particularly in tuberculous cases, it is possible that the high temperature obtained by the Tallerman treatment will give us good results. In fact, the Koch bacillus offers a poor resistance to the action of heat, and the various attempts at the treatment of local tuberculosis by heat have been so encouraging as to merit continuance.

3. In the treatment of sprains, accidents which have become frequent since the wide introduction of the bicycle.

4. In the treatment of certain atonic ulcers, where healing is hindered and retarded by malnutrition of the adjacent and surrounding tissues. The use of heat in the

treatment of such cases has been tried elsewhere, and certain patients suffering from varicose ulcers have been benefited by the application of compresses soaked in water almost at the boiling-point. We have read of some English cases in which chronic ulcers have been treated by the Tallerman method with a success which we, unhappily, have been unable to attain.*

5. Thanks to Mr. Tallerman's apparatus for the pelvic region, lumbago can be treated as well as the diseases of the hip and abdomen (coxalgia, sacrocoxalgia), (Brodie's disease).

6. I mention here hysterical coxalgia, for I believe that these hot-air baths can with benefit be added to the means which have been used to treat the painful or paralytic symptoms, both motor and sensory, of hysteria.

7. In conclusion, there is a disease with which a few experiments will prove interesting. I refer to soft chancre.

If it is true that the Ducrey-Unna bacillus succumbs to high temperatures, it is possible that we may obtain results with the Tallerman apparatus which have not yet been obtained in the treatment of soft chancre by prolonged hot baths, hot applications, etc.

This is only a simple hypothesis, which it will be easy to prove. The apparatus we have hitherto used allowed us to treat the limbs or parts of the limbs only. Thanks to the pelvic apparatus Mr. Tallerman has furnished to us, into which the whole abdomen can be introduced, it will be easy to experiment on soft chancres of the genital organs.

To sum up, the results obtained by the use of the Tallerman apparatus, both in England, where it is in everyday use in private and hospital practice, and in Paris, have demonstrated a real progress in the therapeutic employment of hot air.

Although the list of diseases treated by this local application of superheated dry air appears somewhat long and diverse, we must say that there is no intention to claim that this treatment will cure all diseases. Such a pretension has never been made.

° See photos of cases since treated.

All that we can say is, that it seems to be destined to render great service in the hands of the physician and surgeon, for, on the one hand, it has taken effect on diseases reputed to be incurable, such as chronic, deformatory rheumatism, against which the physician up till now confessed himself to be completely powerless; and, on the other hand, it has acted more quickly and with better results than all other therapeutic methods in diseases often intractable, such as sciatica, gout and gonorrhœal rheumatism.*

VI.

Report from the North-West London Hospital.

The authorities of the North-West London Hospital having, at the suggestion of Mr. Mayo Collier, kindly consented to submit a series of cases to treatment by the localized superheated dry-air bath, the proprietors of that apparatus courteously placed one of their cylinders at the disposal of the medical staff for that purpose, and I give the notes and results obtained during the past two months. It might not be out of place to refer here to the fact that this apparatus was introduced to the medical profession by Mr. Willett at St. Bartholomew's Hospital, in the wards of which institution a variety of cases were treated for two months, after which Mr. Willett detailed the results obtained in a clinical lecture delivered on May 23 last. Acting upon the suggestions thrown out by the lecturer on that occasion, several medical cases were treated. The operation extends over forty minutes, at an average temperature of from 240° to 260° F., and may be shortly described as follows: The affected part is placed in the cylinder, which, to save time, has been already heated to a temperature of 150°, and is then gradually raised. The system of heating and ventilation admits of the air in the cylinder being kept

* From *La Presse Médicale*, December 26, 1896.

practically dry throughout the operation, thereby enabling a very high temperature to be borne by the patient, in one instance at St. Bartholomew's Hospital reaching 300°, whilst temperatures from 270° to 280° are by no means unusual. Under treatment patients experience a sense of comfort, probably due to the high temperature exercising an anodyne influence, which relieves the pain, or more often removes it entirely; even when adhesions have been broken down, the pain is much modified if the joint is immediately subjected to treatment. Some of the cases can hardly fail to be of interest, being of that chronic character for which so little can be done by ordinary medical treatment; all of them were selected for their severity in order to test to the utmost the value of the apparatus. The majority were cured, and the remainder exhibited such marked improvement, that it is only fair to state that there was not a single failure. Mr. Lewis A. Tallerman kindly gave his personal attention during the treatment, and Mr. Mayo Collier supervised the selection of the cases.

J. F. SARGEANT, M.R.C.S., L.R.C.P.,
Resident Medical Officer.

Editorial Comments by the 'Lancet.'—It must be confessed that the results obtained by the usual treatment in cases of chronic stiffness of joints are far from satisfactory, so that one is inclined to welcome all the more cordially a recent therapeutic method which claims, and apparently with justice, to remove, at least in part, this reproach from surgery. The Tallerman Local Dry Hot-Air Bath is an invention by which dry air at temperatures from 250° to 300° F. is applied to a portion of the body, such as hand or foot, knee or elbow, an arm or leg. In May of last year Mr. Alfred Willett delivered a clinical lecture at St. Bartholomew's Hospital on some cases which he had treated by this method; and the present list consists of a series of eight cases in which this bath was employed at the North-West London Hospital, under the care of Mr. Mayo Collier. The bath itself consists of a copper cylinder, varying in size and shape according to the part to be enclosed; and instant relief can be afforded the patient if the temperature causes any discomfort. The cylinder is heated by gas-burners placed underneath. Precautions are taken to prevent the skin from coming into contact with the heated metal. Mr. Willett's conclusions are decidedly in favour of the treatment in certain cases, and his results may be summarized as follows: The first effect of the heat is to induce a copious diaphoresis, and the circulation of the blood in the part is enormously increased, as is well shown by the bright redness of the limb when removed from the bath. The anodyne effect is often remarkable; pain is generally not only relieved, but entirely removed, so that the patient expresses the great relief he feels, and moves the limb with much greater freedom and with much less pain. The cases that appear the most likely to be relieved by the treatment are sprains, stiff joints (where there are no very strong adhesions), flat-foot, gonorrhœal rheumatism, acute and chronic gout, chronic ulcers, and rheumatism. It deserves also to be employed for its anodyne effect after forcibly breaking down under an anæsthetic adhesions which have formed in or around a joint; if the limb be placed in the cylinder, the pain, which is generally severe, is greatly lessened, and the secondary stiffness is much diminished. Mr. Willett was of opinion that but little assistance would be afforded by this bath in overcoming firm, fibrous, articular adhesions. 'Sooner or later the adhesions will have to be forcibly broken down under an anæsthetic, but, such having been done, I think the recovery will in many cases be hastened by the subsequent use of the heated cylinder.' In other words, while facility in active movements is immensely increased, the range of movement is not extended to any great amount. The results obtained in the cases treated at the North-West London Hospital endorse the favourable opinion of Mr. Willett; and the benefit to be derived in rheumatic arthritis is shown in two cases, and in gout in one. The general effect of the baths is marked, but apparently in no way injurious. The entire skin is relaxed and perspires freely, and pain is relieved in joints other than those enclosed in the cylinder.*

VII.

Report of Meeting of Harveian Society.

A meeting of this society was held on May 21, Dr. Gow being in the chair.

Dr. Knowsley Sibley showed two cases of rheumatoid arthritis successfully treated by the hot-air method. The first, a woman 66 years of age, came under his care at the North-West London Hospital in 1892, having previously been under treatment at another hospital. At this time she had considerable enlargement of all the fingers of both hands. These were quite fixed and very painful. The movements of the wrists were also very limited, so also was the right elbow. The knees and shoulders were also enlarged and painful. The patient attended for many months, and, although a large number of drugs and remedies were tried, she gradually became worse, and had to give up her occupation. Then she was unable to dress, and finally unable to feed herself, and was becoming more hopeless every day. At this time—August 1, 1894—one of the Tallerman superheated dry-air apparatus was brought to the hospital, and this new treatment was prescribed. After the first application the pain was considerably relieved and the joints showed some improvement. After the fourth application the patient, who had been unable to follow her business for many months, was again able to use her needle. After the eighth application, with the exception of the right index-finger, all the others could be flexed without much difficulty. On August 24, after the ninth operation, the patient reported she had resumed her occupation, and she was able to walk up and down stairs

* From the *Lancet*, January 12, 1895.

without pain. The patient continued under this treatment till September 13, having then had twenty baths, when she considered herself practically cured; but as the apparatus was still at the hospital, she had an occasional bath up to the end of October, 1894. From that date to November, 1895, she continued at her work, and, with the exception of minor ailments, suffered no inconvenience from her former trouble. At this time she again complained of pain and stiffness in the right hand, so three more baths were prescribed; she again rapidly improved, and from that time to the present has continued much as she now is, and has not relapsed. The other case was a woman aged 69 years. For some years her hands had been painful and the joints enlarged. She first came under Dr. Sibley's care at the hospital on April 22, 1896. Her knees and ankles were somewhat stiff and painful, but the hands were chiefly affected; all the finger-joints were much enlarged; she also suffered much from bronchitis and asthma. This patient had now had fifteen baths, with the result that her knees and ankles were much less painful and her hands much less swollen; so also her cough and general nervous condition had greatly improved. Dr. Sibley explained that the treatment consisted in placing one limb, usually the most affected one, in a cylinder heated to about 150° F., and this temperature was gradually increased up to about 240°. The limb was allowed to remain in for about forty minutes. During this time the patient usually perspired freely from the limb, and, in fact, from the whole body. The temperature as taken in the mouth rises one or two degrees, and the frequency of the pulse is also increased. About a quarter of an hour after the limb is removed the temperature returns to what it was before the treatment. The pain in the affected and other parts was greatly relieved, and the patient experienced a considerable feeling of relief generally; especially it was noticed that the bronchitic condition, which so often accompanies this affection, was also much benefited.*

* From the *Lancet*, June 6, 1896.

VIII.

Report of Meeting of Philadelphia County Medical Society

The President, Dr. J. C. Wilson, was in the chair.

Dr. Horatio C. Wood presented to the society Mr. Lewis A. Tallerman, who came from London with a letter of introduction from Dr. Lauder Brunton. Mr. Tallerman delivered a short address on the Tallerman Patent Superheated Dry-Air Apparatus for the localized application of dry air heated to a high temperature.

The efficacy of the apparatus was tested on two patients from the Philadelphia Hospital—one a case of saturnine gout, the other one of lumbago. Dr. H. C. Wood said that Mr. Tallerman is the inventor of a new method of treating chronic rheumatism, rheumatoid arthritis, sprains, both acute and chronic, and a large number of general inflammations and rheumatic affections. There have been published in the journals, especially in the *Lancet*, a number of very remarkable results, said to have been obtained by the use of the apparatus of Mr. Tallerman, who will briefly state the principles of its construction, and show its immediate mechanism, and then it is proposed to put the apparatus to the test on two patients whose treatment will require from forty to sixty minutes, so that the results of the treatment can be seen.

Dr. Frederick A. Packard demonstrated a case of saturnine gout that had been extremely obstinate to all forms of treatment, in so far as his auricular symptoms were concerned, and had at the time suffered from severe pain in one of the big toes, which was constant in spite of all that could be done. On learning of the Tallerman method of treatment, Dr. Packard thought that the case would be a good one for demonstration, because the man was suffering pain in the big toe so that it could not be touched or flexed, and also because he had tender nodes on the hands, so that the effects on peripheral processes far from the seat of treatment could be seen.

Dr. H. C. Wood, in answer to the question, 'Is not moisture produced by the great heat and liability of scalding present?' said that the only moisture about the apparatus is that from the limb, and this is said to be got rid of. He presented a case of lumbago, in which it was supposed the apparatus would afford marked relief.

After the demonstration Dr. Wood pointed out that the man with the lumbago, who had moaned and groaned when touched, and had been in the hospital a week without much benefit, was, after the treatment, able to get up.

The case of saturnine gout had been under competent medical care for some weeks, without much gain. The man had not moved his toes for three months, but after the treatment he was able to move the toes freely. In fifteen minutes after the treatment was begun he could move his toe, while the pulse went up twenty beats and the temperature rose $1 \cdot 4°$; the other man's temperature went up about $1°$ in forty minutes, and his pulse increased nine beats and gained very distinctly in fulness. The temperature within the apparatus was $260°$, and the pelvis was kept in it for forty minutes. The patient with saturnine gout stood a temperature of $260°$ for three minutes; then it had to be reduced to $255°$, which he stood for some time. Most of the time the temperature was $248°$.

Dr. H. C. Wood said that the immediate results achieved by the apparatus were remarkable, and that it is stated on the highest authority in England that permanent results are reached; so that the apparatus seems to be an important addition to our therapeutics in a class of cases that have been exceedingly anxious and worrying to all.

IX.

**Paper by Dr. Knowsley Sibley, in the 'Lancet,'
August 29, 1896.**

Physicians have shown a marked tendency of late to treat many complaints by what may be termed local or external methods, in contradistinction to treatment by the exhibition of drugs and internal remedies. Even many cases of heart disease are now much improved, if not occasionally cured, by regulated exercises and baths (Schott treatment). In many cases this improvement has taken place without the assistance of drugs of any kind. Undoubtedly many heart cases which had failed to show any permanent benefit by a carefully regulated course of internal medicines have been greatly improved by the Schott treatment. It is not wished in any way to depreciate the value of internal remedies, which still prove of great use when combined with external methods; but the object of this paper is to emphasize the fact that the external method is often the most powerful and effective means of dealing with disease. Rheumatism in its various forms has from times of ancient history been treated by external applications, such as blisters, cupping, leeches, hot sand, friction, warmth, and more recently by massage, electricity, and the like. A large number of baths and moist hot-air methods are at present in vogue; in fact, every fashionable health resort has its own particular kind of charm in the form of a bath, with or without an electrical current. But the one feature which at once appears fatal to satisfactory results is that in all forms a moist and not a dry heat is applied to the affected part. As a result, it is impossible to obtain a really high temperature, as moist heat of from 115° to 120° F. is unbearable, and much above this produces scalds. So with all these older methods, what seems to be the important element in the treatment—namely, heat of from 200° to 300°—cannot be applied, or, at any rate, not for a sufficient length of time.

The method about to be described was first brought

under my notice in August, 1894, when an apparatus was supplied to the North-West London Hospital and left for the use of the staff. I have tried the treatment in a large variety of cases ever since, and now wish to publish some of the results. This period of two years has enabled me to form a definite opinion that the value of the treatment is not merely a temporary one, but is of a more or less permanent nature. The patient, suitably clad in flannel to encourage free perspiration and prevent too much radiation of heat from the body, whether seated on a chair or lying in bed, suffers no inconvenience or discomfort from the high temperature. The treatment lasts usually from fifty minutes to an hour, and it is to this prolonged application that the therapeutic effect to be described is, to my mind, mainly, if not solely, due. When the part is first put into the chamber the temperature is usually about 150°, and this is gradually raised to 220°, and thence upwards, in some cases to as high as 300°. When the treatment is required to act quickly as an anodyne, the temperature is rapidly raised to 260° or 280°. But under ordinary circumstances, such as those described below, it is gradually raised, and a general free perspiration breaks out over the whole body; at the same time the body temperature is temporarily raised from a half to three degrees, a physiological effect hitherto regarded as impossible to be obtained. Also, the pulse increases in frequency and to a less marked extent the respiration. A few minutes after the operation is completed the pulse, respiration, and temperature return to the normal or previous condition. Usually about an hour after the pulse is found to be slower and stronger than it was before treatment; this was especially noticed in some cases of weak and enfeebled hearts. In cases accompanied with much pain this is almost at once relieved, and under the influence of the heat the parts soon become more lax and supple. When the limb is first removed there is often a transient erythematous blush. After the bath the whole body is briskly and lightly rubbed down with a dry towel, and the limb sometimes gently massaged with olive or other oil. The patient then waits until quite cool before going out of the room in order to avoid the risk of a chill.

It will be seen that this method differs materially in the following points from a Turkish bath : (1) The temperature is higher. (2) The application of this temperature is continuous for nearly or quite an hour. (3) The patient breathes the air of the room, and not that of the heated chamber. (4) The application of the heat is only local, the most affected part being treated. It may here be remarked that this local treatment has a general effect, and it is evidenced by the result that, although the particular part treated receives the greatest benefit, other parts of the body affected, but not actually immersed in the chamber, also much improve both with regard to pain and to increase of movement. (5) The treatment does not tend to produce cardiac depression even in the very feeble, or those conditions of exhaustion which are in some cases apt to occur when heat is applied to the whole surface of the body and at the same time inspired. A Turkish bath is less stimulating in its effects, and this local treatment may be confidently recommended in cases in which the former would be quite out of the question. (6) The portable character of the apparatus enables it to be taken to the sick-room and used by the bedside in cases where it would be impossible to move the patient. (7) The local bath gives far more successful results as a method of treatment.

* * * * *

It must be admitted by those who have had much practical experience of severe cases of arthritis deformans how very hopeless these have generally been considered, and what little chance of permanent good medical treatment offers, that this treatment by hot air at a very high temperature meets a general want. I must add I have never seen results so immediate and satisfactory produced by any other treatment. It is now two years since treatment by the local hot-air bath was commenced in the first case, and yet the patient continues comparatively free from the complaint, and even the deformity of the fingers has greatly disappeared. Very intractable cases of sciatica usually pass into the hands of the surgeon, who performs nerve stretching, often with very little good result. It seems likely that these cases will likewise in future be cured by this less drastic means. Another possibility of much importance may be the prevention of many cases of the morphia habit. Judging from the general physical improvement that all of the patients showed after undergoing treatment, it would appear that this method will be found beneficial for cases of chronic bronchitis and for some cases of chronic heart mischief. Many of the patients have stated that their bronchitis was better than it had been for years. I possess also evidence of the undoubted value of this hot-air treatment in acute and chronic gout, and hope shortly to publish a series of cases.

With regard to the physiology of the processes. Locally (1), the heat produces dilatation of all the cutaneous vessels and free circulation through the parts—it is impossible to say how deeply into the tissues this extends, but from the results it may be judged to be some distance—and at the same time there is a marked stimulation of the nutrition of the cutaneous nerves; (2) there is free perspiration of an acid sweat; and (3) relief from pain, however produced, is almost at once apparent. Generally (1), there is profuse perspiration and dilatation of vessels; (2) increase of the rate of the pulse and force of the heart's action; (3) increase (slight) of the respiratory movements; and (4) an increase in the body temperature often of 2° or 3° F. The treatment appears to lower the blood pressure of the body, and in some way to increase the alkalinity of the blood, which enables it to dissolve the uric acid from the tissues and joints and get rid of this substance through the various excretory organs. This is evidenced by the relief from local pain and the removal of the frequent uric acid nerve depression. Hence the treatment is of a tonic nature, and bestows an increased general vitality upon the patient.

So far, although it is now two years since this new treatment was brought under my notice and tried by myself, I am not cognizant of any unsatisfactory result following it, notwithstanding the fact that many of the cases have not been very promising ones to do much with; many were old and debilitated people, and in some cases the objective disease was attended by heart and other visceral complications. And what is of all importance with regard to satisfactory treatment, the patients were in every case treated as out-patients, and so there was no control over their habits. In a clinical lecture delivered at St. Bartholomew's Hospital, in which this new treatment was first introduced to the profession in this country, the lecturer described it as one that was 'not only quite painless,' but might be 'almost called that of luxurious ease.' My experience fully confirms the soothing effect in this process of relieving disease accompanied with pain. Patients worn out through prolonged suffering will frequently, if permitted, fall asleep whilst under the operation. One of the most important therapeutic effects is the sleep which usually follows the treatment, particularly in cases of patients who have been prevented by pain from getting any rest for long periods.

It does not come within the province of this article to enter into the various surgical conditions in which this treatment has been found efficacious. Papers on this subject have already been published by Mr. Willett in the above-mentioned lecture; in 'The Deformities of the Foot,' by Mr. W. J. Walsham; in an article by the same writer in the Transactions of the American Orthopædic Association, 1895; and also reports of cases treated at the North-West London Hospital, which were published in the *Lancet* of January 12, 1895. At the present time the above cases may be of interest to those engaged in this treatment and to those anxious to try it. Undoubtedly the writer's experience points to the conclusion that the chronicity of a case is by no means a bar to a successful issue. With regard to rheumatoid arthritis, many forms of rheumatism with chronic joint mischief, sciatica, lumbago, and, one might add, some cases of neuralgia, neuritis, and chronic bronchitis, there can be no question as to the great benefit of the treatment. The record of the remarkable results previously published obtained by this apparatus at the North-West London Hospital alone embraces such a wide field that this localized hot-air bath must be reckoned upon in the future to play an important part in the relief of pain and the cure of disease. Especially is it likely to prove of great use in those forms of very chronic disease which have hitherto yielded but little to any known medicine, the sufferers from which are commonly sent in search of Continental hydrothermic establishments, and usually, sooner or later, fall into the hands of the quack.

X.

Report from the Liverpool Workhouse Infirmary from July, 1896, to February, 1897.

During this period fifty patients underwent the localized hot-air treatment, twenty-eight of them for surgical complaints and twenty-two for medical. Of these fifty patients, sixteen were cured, eighteen were improved, fourteen derived no benefit, and the remaining two were unable to continue the treatment owing to faintness. Of the eighteen who were improved, five were considerably relieved and thirteen only slightly so; two of the latter, however, are still undergoing treatment, and express themselves improving with each successive bath.

Of the sixteen patients who were cured, eight were treated from the surgical wards and eight from the medical wards. Of those who were improved, nine were treated for medical and nine for surgical complaints. Of the fourteen who derived no benefit from the treatment, four were from the medical wards and ten from the surgical wards. Of those ten surgical cases two were ulcerated, indolent, sore legs, which had resisted all previous treatment, and were therefore tried with the bath. Strange to say, both of these patients after the ninth or tenth bath became very exhausted, and one developed gastric symptoms of a rather serious character. Of the two who were unable to continue the treatment, one was a surgical, the other a medical patient.

All the cases treated from the surgical wards were, with a few trifling exceptions, in the subacute stage, while those from the medical were all, or almost all, of the chronic type.

With regard to measurements, it has been found that out of the fifty cases treated, a decrease in the size of the affected part (the difference of measurement being found between the part affected previous to the first bath and after the last) has taken place in twenty-one instances, no change in thirteen, and an actual increase in nine; of the remaining seven, no notes were taken of measurements in three cases, the two ulcerated legs and another two are still undergoing treatment, and two were unable to bear the process of treatment.

Out of the twenty-one cases where there was a decrease of the affected part, nine derived little or no benefit, while twelve were greatly improved or cured by the bath. Out of the thirteen where no change of measurement in the affected part was brought about by the bath, ten derived little or no benefit, while three were greatly improved or cured. Where there was an actual increase of measurement after treatment by the bath, it has been found that four were not improved much, if at all, while three were either greatly improved or cured.

XI.

Report from the Livingstone Cottage Hospital, Dartford.

By THOS. F. CLARKE, M.R.C.S., Etc.

The first two of the following cases (omitted here) were submitted by me to Mr. Tallerman for treatment at a demonstration of the Tallerman Local and Superheated Dry-Air Bath, given at the Livingstone Cottage Hospital, Dartford, on March 18, 1896.

Mr. Tallerman afterwards kindly allowed the cylinder to remain at the hospital for a month, in order that further opportunities might be afforded of testing the therapeutic value of that apparatus in the continued treatment of these and such other cases as might be considered suitable.

From the observations I have made of this form of treatment, I have found that, as a general rule, far better results are obtained when the limb directly operated upon perspires freely, and at the same time there is general diaphoresis. This does not occur in all cases (owing to the inactive, if not unhealthy, condition of the skin), until the patient has been treated on several occasions. It did not occur with the first case, and although, as I have subsequently remarked, this was an extremely severe test-case for illustration, I believe still greater improvement would have followed as soon as free diaphoresis had been attained.

One very noticeable feature during the application of the hot-air bath was *the effect produced on the heart: its action, although slightly accelerated as the temperature of the bath is raised, is rendered considerably stronger, and the pulse becomes much firmer and fuller. This result is especially of importance, as showing that patients in a very debilitated* condition may with safety be subjected to this form of treatment.

There is no doubt a great future in store for this mode of treating many cases, as its value has been amply proved in such cases as acute inflammation of joints, severe sprains, gout, many rheumatic affections, etc.

       (Signed) THOMAS F. CLARKE.
April 29, 1896.

N.B.—Since writing the above, I have had opportunities of seeing all the patients whose cases I have recorded here. In no single instance have the good results afforded by this treatment disappeared, thus proving that the hot-air bath exerts a more or less permanent effect.

September, 1896.

XII.

Letter to the 'Lancet' from F. Fitzherbert Jay, M.D.

To the Editor of the ' Lancet.'

SIR,

  Referring to the report of cases by Dr. W. K. Sibley in the *Lancet* of August 29 last which were treated by the above method, I am glad to be able to confirm from personal experience the curative effect of a course of only four baths for chronic sciatica and lumbago, which had begun to extend to the sound leg. In the severe winter of 1894 I had an attack of acute sciatica, which has continued in a chronic form. Having acted as resident physician to three hydropathic establishments, I had enjoyed considerable experience in the treatment of such cases, and was enabled to apply the knowledge so acquired to my own case. The relief obtained was but slight, and I had difficulty in getting about, the pain at times being very severe. I am glad to be able to record that I derived benefit from the first hot-air bath of the Tallerman process. Having now taken four, I am free from pain, and able to walk three miles at a stretch. No ill effects or risk of any sort are likely to attend a course of these baths, a slight rise of temperature and increase of pulse being alone indicated. The baths were taken at the Institution, 50, Welbeck Street, W., and I am indebted to Mr. Lewis A. Tallerman for much kindness and attention.

I am, sir, yours faithfully,
(Signed) F. FITZHERBERT JAY, M.D., St. And.

CONSTITUTIONAL CLUB,
October 7, 1896.

APPENDIX.

NOTE BY THE INVENTOR.

The Tallerman Treatment Institute,
50, Welbeck Street,
Cavendish Square, London.
November 9, 1897.

In answer to the many inquiries and statements which have been made with respect to the origin and invention of the 'Tallerman Treatment,' it may be stated that the facts connected therewith are as follows, viz.:

In May, 1893, I was asked by Mrs. E. D. T. Sheffield and Mr. T. H. Rees to take an interest in an apparatus for treating stiff joints by *superheated steam*, a method by which it was stated a limb could be treated at from 250° to 300° F. On the 30th of that month an agreement was entered into to test the efficiency of the apparatus, and binding the parties to patent in the joint names any improvement or new invention any one of them might devise for a similar purpose.

The investigation commenced by the treatment of my own foot in the apparatus, a cylindrical-shaped Receiver, the distal end of which was closed and steam-tight. Steam was generated in a boiler communicating with the interior of the Receiver by a pipe, through which I was informed superheated steam was passing under pressure, and that the temperature at which my foot was then being treated was 280° F. The absence of any sensation of discomfort or pressure which steam, superheated to such a temperature, might be expected to produce appeared very remarkable. Further investigation was postponed, to be continued at

Harrogate about a month later, when, thanks to the introduction of Dr. Lever, the staff of the Bath Hospital were good enough to consent to the apparatus being tested at that institution. The net result of the investigation was :

1. To demonstrate that the impression as to the high temperature obtainable was clearly erroneous. In the presence of the staff and committee, a thermometer registering up to 212° F. only was placed in the Receiver. At the expiration of about an hour, Mr. Rees expressed himself satisfied that he could not raise the temperature in the Receiver any higher. The thermometer was then examined and found to register 160° F. only.

2. It was found that even this temperature soon became unbearable to the patient, and either the limb was removed or a reduction in the temperature was effected by sponging the outside of the Receiver with cold water.

3. Further, it was elicited that the latter measure led to condensation of the vapour, and that the hot water at times fell on the enclosed limb. In this way Mr. Rees had himself been scalded and also a patient.

4. That the apparatus was, in fact, a local-vapour bath, that it did not carry out the object for which it had been devised, and that its use was accompanied by considerable risk.

The apparatus and boiler were forthwith returned to London, and this method of treatment was never afterwards used. Meanwhile, becoming greatly interested in the problem of treating disease by heat, I turned my attention to this branch of therapeutics, and soon became convinced that a *very high temperature* could not fail to exercise a markedly beneficial effect in various diseases if it could be applied *without pressure, and continuously over a sufficient length of time on each occasion.*

Applying myself to the subject, I succeeded in inventing the present method known as the 'Tallerman treatment,' and the contents of this volume show my views were not incorrect. It will be noted that a patient was treated at St. Bartholomew's Hospital at a temperature of 300° F.

(see Clinical Lecture, *Clinical Journal* of May 30, 1894); two patients were treated at the Philadelphia Hospital at 310° and 315° F. respectively (see Notes); and it might be stated that with regard to duration a patient has been treated under medical observation for upwards of two hours at a temperature varying from 260° to 270° F. In all the foregoing cases the effect was soothing throughout; in the last there was lassitude on the next day, which was not unexpected, but that was the only discomfort.

The evidence of the value of this new treatment in most intractable and incurable diseases will be found in the Reports and Case Notes; and that a large variety of cases in which the treatment up till now has not been used will benefit by it considerably may confidently be anticipated by anyone who will take the trouble to consider the physiological and therapeutic effects it produces.

In accordance with the before-mentioned agreement, the system and apparatus was patented in the joint names, but is known as the 'Tallerman treatment,' and has been in use since.

I think it only right to avail myself of this opportunity to express my acknowledgments to the distinguished members of the medical profession whose names appear in this book for the pains they have bestowed on the investigation of my invention, and for the kind terms in which they have recognised its merits. Had it not been for their friendly encouragement, the prejudice I have encountered in other quarters would have long ago compelled me to give the thing up, and allow it to fall into the hands of those who would have worked it as a commercial concern independently, and to the detriment of the medical profession.

LEWIS A. TALLERMAN.

www.ingramcontent.com/pod-product-compliance
Lightning Source LLC
Chambersburg PA
CBHW022117230426
43672CB00008B/1416